Basic Electromagnetism

PHYSICS AND ITS APPLICATIONS

Series Editors

This series of short texts on advanced topics for students, scientists and engineers will appeal to readers seeking to broaden their knowledge of the physics underlying modern technology.

Each text provides a concise review of the fundamental physics and current developments in the area, with references to treatises and the primary literature to facilitate further study, Additionally texts providing a core course in physics are included to form a ready reference collection.

The rapid pace of technological change today is based on the most recent scientific advances. This series is, therefore, particularly suitable for those engaged in research and development, who frequently require a rapid summary of another topic in physics or a new application of physical principles in their work. Many of the texts will also be suitable for final year undergraduate and postgraduate courses.

1. **Electrons in Metals and Semiconductors**
   R.G. Chambers
2. **Basic Digital Electronics**
   J.A. Strong
3. **AC and DC Network Theory**
   Anthony J. Pointon and Harry M. Howarth
4. **Nuclear and Particle Physics**
   R.J. Blin-Stoyle
5. **Thermal Physics**
   Second edition
   C.B.P. Finn
6. **Vacuum Physics and Techniques**
   T.A. Delchar
7. **Basic Electromagnetism**
   E.R. Dobbs

Basic Electromagnetism

E. R. Dobbs

Emeritus Professor of Physics
University of London

CHAPMAN & HALL

London · Glasgow · New York · Tokyo · Melbourne · Madras

Published by Chapman & Hall, 2–6 Boundary Row, London SE1 8HN

Chapman & Hall, 2–6 Boundary Row, London SE1 8HN, UK

Blackie Academic & Professional, Wester Cleddens Road, Bishopbriggs, Glasgow G64 2NZ, UK

Chapman & Hall Inc., 29 West 35th Street, New York NY10001, USA

Chapman & Hall, Japan, Thomson Publishing Japan, Hirakawacho Nemoto Building, 6F, 1-7-11 Hirakawa-cho, Chiyoda-ku, Tokyo 102, Japan

Chapman & Hall Australia, Thomas Nelson Australia, 102. Dodds Street, South Melbourne, Victoria 3205, Australia

Chapman & Hall India, R. Seshadri, 32 Second Main Road, CIT East, Madras 600 035, India

First edition 1993

Typeset in 10/12 Times by Thomson Press (India) Ltd, New Delhi
Printed in Great Britain by St. Edmundsbury Press, Bury St. Edmunds, Suffolk

ISBN 0 412 55570 0

A Catalogue record for this book is available from the British Library

Library of Congress Cataloging-in-Publication data

Dobbs, Roland, 1924–
Basic electromagnetism / E. R. Dobbs.—1st ed.
p. cm.—(Physics and its applications ; 7)
Includes index.
ISBN 0-412-55570-0
1. Electromagnetism. I. Title. II. Series.
QC760.D62 1993 93-7067
537—dc20 CIP

Printed on permanent acid-free text paper, manufactured in accordance with the proposed ANSI/NISO Z 39.48-199X and ANSI Z 39.48-1984

Contents

Preface

Electricity and magnetism are basic to our understanding of the properties of matter and yet are often regarded as the difficult parts of an undergraduate course in physics, materials science or engineering. In the first six chapters of this book answers are developed from first principles to such questions as: What is electricity? What is electromagnetism? Why are some materials magnetic and others non-magnetic? What *is* magnetism?

These questions can be answered in two related ways. On the one hand the *classical* explanation is in terms of classical concepts: electric charge (q), electric and magnetic fields ($\boldsymbol{E}$ and $\boldsymbol{B}$) and electric currents. On the other hand the *microscopic* (or 'atomic') explanation is in terms of quantum concepts: electrons, nuclei, electron orbits in atoms, electron spin and photons. Microscopic explanations underlie classical ones, but they do not deny them. The great triumphs of classical theory are mechanics, gravitation, thermodynamics, electromagnetism and relativity.

Historically the classical theories began at the time of Newton (seventeenth century) and were completed by Maxwell (nineteenth century) and by Einstein (early twentieth century). Microscopic explanations began with J.J. Thomson's discovery of the electron in 1897. For most physical phenomena it is best to seek a classical explanation first, especially for phenomena at room temperature, or low energy, when quantum effects are small. This book presents classical theory in a logical, self-consistent sequence, but reference is made to microscopic (quantum) theory at each appropriate stage.

Electromagnetism began in 1819 with the discovery by Oersted that an electric current is associated with a magnetic field and was followed in 1820 by Ampère's discovery that two wires carrying electric currents exerted magnetic forces on one another. But it was Faraday's discovery of electromagnetic induction in 1831, or as he put it, the conversion of magnetism into electricity, that finally showed that electricity and magnetism were not distinct, separate phenomena, but interacted when there were time-varying electric or magnetic fields. The beauty of electromagnetism is that Faraday's experiments led to a symmary of the whole of electromagnetism in just the four equations of Maxwell's theory, which relate $\boldsymbol{E}$ and $\boldsymbol{B}$ in space with fixed and moving charges, together with the electromagnetic force law. These equations are greatly simplified when we deal with *statics*, that is, variables that do not depend on time t, or *stationary* variables. Maxwell's equations then simplify and separate into two independent pairs of equations:

1. The first pair describe the electrostatic field $\boldsymbol{E}$ for fixed charges and are known as *Gauss's law* and the *circulation law*. They summarize electrostatics.
2. The second pair describe the magnetostatic field $\boldsymbol{B}$ for steady currents (charges moving at constant speed) and are known as *Gauss's law* and *Ampère's law*. They summarize magnetostatics.

In electrostatics only the $\boldsymbol{E}$ field appears and $\partial E/\partial t = 0$; in magnetostatics only the $\boldsymbol{B}$ field appears and $\partial B/\partial t = 0$. So under these conditions, electricity and magnetism *are* classically distinct, separate phenomena. But if you charge a capacitor (q varying with time) or move a magnet ($\boldsymbol{B}$ at a point varying in time) then $\boldsymbol{E}$ and $\boldsymbol{B}$ are no longer independent and new terms in the equations due to *electromagnetism* appear, as first discovered by Faraday (*Faraday's law*) and Maxwell (*Maxwell's law*).

The development of the subject in this text is therefore first electrostatics, then magnetostatics, followed by electromagnetism and magnetism. The seventh chapter summarizes electromagnetism in terms of Maxwell's equations, which are then used to study the propagation and generation of electromagnetic waves. The first seven chapters often comprise a first course in electromagnetism for undergraduates.

In studying the solutions of Maxwell's equations you will find answers to such questions as: What is an electromagnetic wave? Why does a radio wave travel through space at the speed of light? How is a radio wave generated? Why does light pass through a straight tunnel when a radio wave does not? How does light travel down a curved glass fibre?

Before studying these solutions, Chapter 8 discusses the remarkable fact that the classical laws of electromagnetism are fully consistent with Einstein's special theory of relativity. The following four chapters provide solutions of Maxwell's equations for the propagation of electromagnetic waves in free space, in dielectrics, across interfaces and in conductors, respectively. In Chapter 13 the generation of radio waves from dipoles and of microwaves from other antennas is explained, while the final chapter shows how these waves can be transmitted down waveguides and coaxial lines. In conclusion, the use of resonant and re-entrant cavities leads to a discussion of the classical theory of radiation and its usefulness as a limiting case of the quantum theory of radiation.

The spectrum of electromagnetic waves covers an enormous range of frequencies, from the very low frequencies (VLF), used to communicate with submerged submarines, to the enormous frequencies (10^{24} hertz) associated with some cosmic rays from outer space. The complete spectrum is illustrated in Appendix D, where it is characterized by both the classical, wave properties of frequency (ν) and wavelength (λ) and the quantized, photon properties of energy ($h\nu$) and temperature ($h\nu/k_B$). Classical electromagnetism provides a theory of the wave properties of radiation over a wide frequency range, including, for example, the diffraction of X-rays by crystals, but for the interactions of radiation with matter classical theory only applies in the long wavelength, low frequency, low energy ($h\nu \ll k_B T$) limit. The generation of electromagnetic radiation is similarly the classical process of acceleration of electrons in producing a radio wave, where the wavelength is macroscopic, but quantum processes are involved in the production of X-rays by electronic transitions in atoms, or gamma rays by nucleonic transitions in nuclei, where the wavelength are microscopic. The production of light by laser action is an interesting example of the combination of the classical process of reflection with the quantum process of stimulated emission. In this book the limits of classical electromagnetism are explained and the usefulness of the wave and particle properties of radiation are discussed, so that the reader is provided with an understanding of the applicability and limitations of classical theory.

The international system of units (SI units) are used throughout and are listed for each electromagnetic quantity in Appendix A. Since Gaussian units are still used in some research papers on electromagnetism, Appendix B lists Maxwell's equations in these units and states the conversion from the Gaussian to the SI system. The physical constants used in the text are listed in Appendix C with their approximate values and units. In Appendices E and F there are summaries of the most useful relations in vector calculus and special relativity. Finally, each chapter, except Chapters 1 and 7, has a set of associated exercises in Appendix G with answers in Appendix H.

ACKNOWLEDGEMENTS

This book owes much to the many undergraduates who have participated in my tutorials and lectures on electromagnetism at the Universities of Cambridge, Lancaster and London for more than twenty years. Of the numerous texts I have consulted, none has been so illuminating as Volume 2 of the *Feynman Lectures on Physics* by R.P. Feynman, R.B. Leighton and M. Sands (Addison-Wesley, London, 1964).

This new, revised edition is a combination of my *Electricity and Magnetism* (Routledge & Kegan Paul, London, 1984) and my *Electromagnetic Waves* (Routledge & Kegan Paul, London, 1985). It is a pleasure to thank colleagues in the Universities of London and Sussex for their helpful comments, especially Professor E.J. Burge for a number of improvements to *Electricity and Magnetism*, after he had used it for a first course to undergraduates in Physics at Royal Holloway, University of London. Thanks are due to the University of London for permission to reproduce some problems (marked *L*) from BSc course unit examinations taken at the end of their first year by students reading Physics at Bedford College.

It is a plesure to thank Mrs Sheila Pearson for her accurate and rapid typing of the original manuscripts and my wife for her constant support and encouragement over many years.

List of symbols

$\boldsymbol{A}$	Magnetic vector potential
A	Area
$\boldsymbol{B}$	Magnetic field
B	Susceptance
C	Closed loop; capacitance
c	Speed of light
$\boldsymbol{D}$	Electric displacement
d	Distance
$\mathscr{E}$	Electromotive force
$\boldsymbol{E}$	Electric field
e	Electronic charge
$\mathscr{F}_{\mathrm{m}}$	Magnetomotive force
$\boldsymbol{F}$	Force
F_{E}	Electrical force
F_{G}	Gravitational force
f	Focal length
G	Gravitational constant; conductance
$\boldsymbol{H}$	Magnetizing field
$\boldsymbol{H}_{\mathrm{m}}$	Demagnetizing field
h	Planck constant
I	Electric current
I_{m}	Surface magnetizing current
$\hat{\mathbf{i}}$	Cartesian unit vector
$\boldsymbol{i}$	Surface current density
i_{f}	Solenoidal surface current density
i_{m}	Magnetization surface current density
$\hat{\mathbf{j}}$	Cartesian unit vector
$\boldsymbol{j}$	Electric current density

$\boldsymbol{j}_{\mathrm{f}}$	Conduction current density
$\boldsymbol{j}_{\mathrm{m}}$	Magnetization current density
$\boldsymbol{j}_{\mathrm{p}}$	Polarization current density
$\hat{\mathbf{k}}$	Cartesian unit vector
$\boldsymbol{k}$	Wave vector
k	Coupling coefficient; wave number
k_{g}	Waveguide wave number
k_{B}	Boltzmann constant
$k_{\mathrm{R}}, k_{\mathrm{I}}$	Real and imaginary parts of $k = k_{\mathrm{R}} - \mathrm{i}k_{\mathrm{I}}$
L	Self inductance
l	Length
$\mathbf{d}\boldsymbol{l}$	Electric current element
$\boldsymbol{M}$	Magnetization
M	mutual inductance
$\boldsymbol{m}$	Magnetic dipole moment
m	Mass
N	Number density; total number of turns
$\boldsymbol{n}$	Unit normal vector
$\hat{n}$	Number of turns per unit length; refractive index
$n_{\mathrm{R}}, n_{\mathrm{I}}$	Real and imaginary parts of $n = n_{\mathrm{R}} - \mathrm{i}n_{\mathrm{I}}$
$\boldsymbol{P}$	Electric polarization
P	Total radiated power
P_{n}	Legendre function
$\boldsymbol{p}$	Electric dipole moment; electron momentum
$\boldsymbol{p}_0$	Permanent electric dipole moment
p_{r}	Radiation pressure
Q	Total electric charge; quadrupole moment; quality factor
q	Electric charge
$\mathscr{R}$	Reluctance
R	Electrical resistance; reflectance
R_0	Reflectance at normal incidence
R_{r}	Radiation resistance
$\boldsymbol{r}$	Distance vector
r	Distance; cylindrical or spherical radius
r_0	Classical electron radius
$\mathscr{S}$	Poynting vector, electromagnetic energy flux
$\boldsymbol{S}$	Surface area vector
$\boldsymbol{s}$	Distance vector
s	Distance; contour
$\boldsymbol{T}$	Torque vector
T	Temperature; transmittance

T_C	Curie temperature
T_F	Fermi temperature
t	Time
U	Energy
u	Energy density; relative velocity
V	Volume; potential difference
$\boldsymbol{v}$	Velocity vector
v_F	Fermi velocity
W	Mechanical work
x	Cartesian coordinate
Y	Admittance
y	Cartesian coordinate
Z	Wave impedance; atomic number
Z_0	Wave impedance of free space
z	Cartesian coordinate
α	Polarizability
β	Absorption coefficient
γ	Electrical conductivity; damping constant
δ	Loss angle; skin depth
ε	Absolute permittivity
ε_0	Electric constant; permittivity of free space
ε_r	Relative permittivity; dielectric constant
$\varepsilon_R, \varepsilon_I$	Real and imaginary parts of $\varepsilon = \varepsilon_R - i\varepsilon_I$
ε_s	Static value of relative permittivity
ε_∞	High frequency limit of relative permittivity
θ	Spherical polar coordinate
θ_c	Critical angle (of incidence)
λ	Electric charge per unit length; wavelength
λ_c	Cut-off wavelength of waveguide
λ_g	Waveguide wavelength of waveguide
λ_0	Free space wavelength of radiation
$\boldsymbol{\mu}$	Magnetic moment of a particle
μ_B	Bohr magneton
μ	Absolute permeability
μ_0	Magnetic constant; permeability of free space
μ_r	Relative permeability
ν	Frequency
ρ	Electric charge density
ρ_f	Free charge density
ρ_p	Polarization charge density
σ	Surface charge density

σ_f	Surface density of free charges
σ_p	Surface density of polarization charges
σ_0	Resonant scattering cross-section
σ_R	Rayleigh scattering cross-section
σ_T	Thomson scattering cross-section
τ	Volume of integration; mean time between collisions
Φ	Magnetic flux
ϕ	Electric (scalar) potential; cylindrical polar coordinate; phase angle of a wave
χ_e	Electric susceptibility
χ_m	Magnetic susceptibility
ψ	Spherical polar coordinate; azimuthal angle; wave function
Ω	General scalar function
$d\Omega$	Element of solid angle
ω	Angular velocity of frequency
ω_c	Cut-off (angular) frequency of waveguide
ω_L	Larmor (angular) frequency
ω_0	Resonant (angular) frequency
ω_p	Plasma (angular) frequency

1

Introduction

Although all electromagnetic phenomena can be studied in empty space, an important part of any introductory course on electricity and magnetism is a proper understanding of the nature of matter. We shall therefore discuss dielectric behaviour in the chapter on electrostatics, conduction in metal wires in that on magnetostatics, and magnetism in matter (whether para-, dia- or ferro-magnetism) in the chapter on magnetism. In this first chapter the nature of matter is summarized.

All matter is composed of elementary particles, some charged positively (protons), some charged negatively (electrons) and some without charge (neutrons). The forces between these particles are of three different sorts – gravitational, electrical and nuclear – which differ enormously in their strength and range.

The *gravitational force* was made famous by Newton in his studies of the planets and expressed by him in 1686 in his law of universal gravitation that 'every particle of matter in the universe attracts every other particle with a force which is directly proportional to the product of the masses of the particles and inversely proportional to the square of the distance between them'. This is the first inverse square law of force which, for two masses m_1 and m_2, is given by

$$F_G = \frac{Gm_1m_2}{r^2} \tag{1.1}$$

where r is the distance between m_1 and m_2 and G is the gravitational constant. The *electrical force* will be familiar as the law Coulomb found in 1785 for the force between electrical charges. This is another inverse square law of force. If r is now the distance between the charges q_1 and q_2, and K is an electrical constant then

$$F_E = \frac{Kq_1q_2}{r^2} \tag{1.2}$$

The third type of force between the elementary particles that constitute matter is a comparatively recent discovery. In 1932 Chadwick found that the nuclei of atoms and molecules contained not only protons but new particles – neutrons – and so there had to be a third type of force, the *nuclear force*, that held these particles together in the nucleus. Collectively the particles in the nucleus are known as nucleons.

This nuclear force, composed of both weak and strong interactions, is exceedingly short range. For example the nuclear force decreases in some cases as $r^{-2}\exp(-r/r_0)$, where r_0 is about $1\,\text{fm} = 10^{-15}\,\text{m}$. An introduction to the exciting properties of nuclei and elementary particles can be found in the book by Professor Blin-Stoyle in this series. In contrast the gravitational and electrical forces are comparatively long range (Fig. 1.1). It is obvious from the motion of the planets round the sun that gravitational forces are long range. It is not so obvious that electrical forces are similarly long range because electrical charges are usually screened by other charges of opposite sign at comparatively short range, so that the overall effect at long range is negligible.

Although both the gravitational and electrical force obey an inverse square law, their strengths differ enormously. For the proton–electron pair which comprises the hydrogen atom, the electrical force F_E is about 10^{39} or one thousand million million million million million million times as strong as the gravitational force F_G, as can

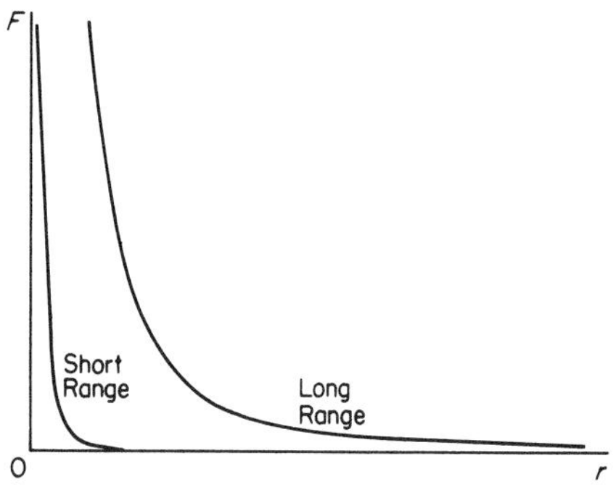

Fig. 1.1 Short-range and long-range forces.

be easily shown from (1.1) and (1.2) and a knowledge of the constants. So we can nearly always neglect gravitational effects in the presence of electrical forces. The exception is experiments like Millikan's oil drop, where the enormous mass of the earth acts on the oil drop with a force comparable to the electrical one exerted on the tiny charges of the oil drop as it moves between the charged plates.

The gravitational and electrical forces also differ in one other important respect: the gravitational force between particles of ordinary matter is always attractive, whereas the electrical force is repulsive (positive) between like charges and attractive (negative) between unlike charges. The net result is that large masses have large gravitational attractions for one another, but normally have negligible electrical forces between them.

The paradox is that although all matter is held together by electrical forces, of the interatomic or intermolecular or chemical-bond types, which are immensely strong forces, large objects are electrically neutral to a very high degree. The electrical balance between the number of protons and electrons is extraordinarily precise in all ordinary objects. To see how exact this balance is, Feynman has calculated that the repulsive force between two people standing at arm's length from each other who each had 1% more electrons than protons in their bodies would be enormous – enough, in fact, to repel a weight equal to that of the entire earth! So matter is electrically neutral because it has a perfect charge balance and this gives solids great stiffness and strength.

The study of electrical forces, electromagnetism, begins with Coulomb's law, (1.2). All matter is held together by the electromagnetic interactions between atoms, between molecules and between cells, although the forces holding molecules and cells together are more complicated than the simple Coulomb interaction. The studies of condensed state physics, of chemistry and of biology are thus all dependent on an understanding of electromagnetism. This text develops the subject from Coulomb's law to Maxwell's equations, which summarize all the properties of the electromagnetic fields, in free space and matter. But if you ask why does the strong electrical attraction between a proton and an electron result in such comparatively large atoms rather than form a small electron–proton pair, you will not find the answer in Maxwell's equations alone. The study of electrical forces between particles at atomic or subatomic distances requires a new physics, quantum mechanics, which is the subject of the book by Professor Davies in this series.

2

Electrostatics

Electric charge has been known since the Greeks first rubbed amber and noticed that it then attracted small objects. Little further progress was made until the eighteenth century when du Fay showed that there were two sorts of charge. One sort followed the rubbing of an amber rod with wool, the other a glass rod with silk. It was Benjamin Franklin who arbitrarily named the latter a positive charge and the original amber one a negative charge. He also showed that the total charge in a rubbing experiment was constant.

2.1 COULOMB'S LAW

In 1785 Coulomb succeeded in discovering the fundamental law of electrostatics. A brilliant experimenter, he was able to invent and build a highly sensitive torsion balance with which he could measure precisely the relative force of repulsion between two light, insulating, pith balls when charged similarly and placed at different distances apart. He showed that this electrostatic force:

1. acts along the line joining the particles;
2. is proportional to the magnitude of each charge; and
3. decreases inversely as the square of the distance apart.

It is therefore a long-range force (Fig. 1.1) and is given by the vector equation for the force $\boldsymbol{F}_1$ on charge q_1 due to charge q_2. Thus,

$$\boldsymbol{F}_1 = \frac{Kq_1q_2}{r_{12}^2}\hat{\mathbf{r}}_{12} \tag{2.1}$$

where $\boldsymbol{r}_{12} = \boldsymbol{r}_1 - \boldsymbol{r}_2$, $r_{12}^2 = (x_1 - x_2)^2 + (y_1 - y_2)^2 + (z_1 - z_2)^2$ and $\hat{\mathbf{r}}_{12}$ is a unit vector drawn *to* 1 *from* 2 (Fig. 2.1) given by $\boldsymbol{r}_{12}/r_{12}$. Fig. 2.1

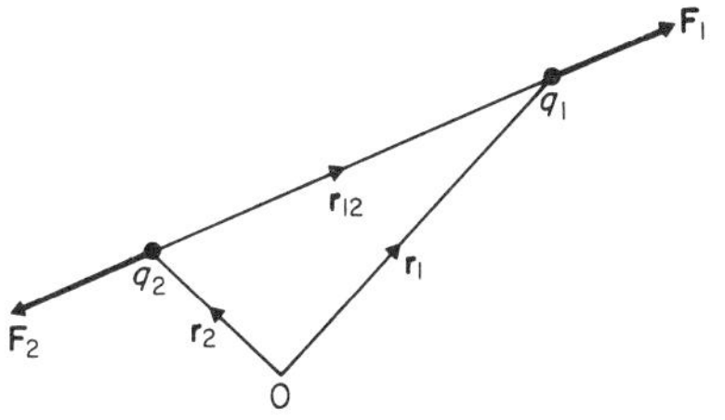

Fig. 2.1 Electrostatic forces between electric charges.

shows that Newton's laws must apply and the force $\boldsymbol{F}_2$ on charge q_2 due to q_1 is $\boldsymbol{F}_2 = -\boldsymbol{F}_1$. When both charges have the same sign, the force acts positively, that is the charges are repelled, while between a negative and a positive charge the force acts negatively and the charges are attracted.

For historical reasons the constant of proportionality K in (2.1) is not one, but is defined as

$$K = \frac{1}{4\pi\varepsilon_0} = 10^{-7}c^2 \tag{2.2}$$

where ε_0 is the *electric constant* (permittivity of free space) and c is the velocity of light. The constant has to be determined from experiment. A recent value of $c = 2.997\,925 \times 10^8\,\mathrm{m\,s^{-1}}$ is accurate to better than 1 in 10^6, but for use in problems can be taken as $3.0 \times 10^8\,\mathrm{m\,s^{-1}}$. On the same basis $K = 9 \times 10^9\,\mathrm{N\,m^2\,C^{-2}}$, using the SI unit coulomb (C) for electric charge.

It is important to note that we have written in (2.1) Coulomb's law for charges in a *vacuum*; we have not mentioned the effects of a dielectric or other medium.

2.1.1 Principle of superposition

The only other basic law in electrostatics is the principle of superposition of electric forces. The principle states that if more than one force acts on a charge, then all the forces on that charge can be added vectorially into a single force. Thus for the total force on a charge q_1 due to charges q_2 at $\boldsymbol{r}_{12}$, q_3 at $\boldsymbol{r}_{13}$, etc., we have:

$$\boldsymbol{F}_1 = \boldsymbol{F}_{12} + \boldsymbol{F}_{13} + \cdots$$

$$F_1 = K\left\{\frac{q_1 q_2}{r_{12}^2}\hat{\mathbf{r}}_{12} + \frac{q_1 q_3}{r_{13}^2}\hat{\mathbf{r}}_{13} + \cdots\right\}$$

$$F_1 = \frac{1}{4\pi\varepsilon_0}\sum_j \frac{q_1 q_i}{r_{1j}^2}\hat{\mathbf{r}}_{1j} \tag{2.3}$$

That the electric force between two small particles can be enormous is readily seen by estimating the force produced in Rutherford's scattering experiment when an alpha particle ($^4_2$He nucleus) makes a direct approach to a gold ($^{197}_{79}$Au) nucleus. The distance of closest approach is 2×10^{-14} m and so the maximum electrostatic repulsion, from (2.1) and (2.2), is

$$F = \frac{2e \times 79e}{4\pi\varepsilon_0(2 \times 10^{-14})^2} = \frac{9 \times 10^9 \times 2 \times 79e^2}{4 \times 10^{-28}} \text{N}$$

Since the charge on the proton, $e = 1.6 \times 10^{-19}$ C, the force on a single nucleus is about 100 newtons and a very strong force.

2.1.2 Electric field

The electric forces due to a distribution of electric charges, and particularly those due to a uniform distribution of charge, are best described in terms of an electric field vector, ***E***, defined as the electric force per unit charge at a point. It can be visualized as the total force on a positive test charge, q at r_1, the magnitude of which is then allowed to become very small so as not to disturb the electric field. Using (2.3) we have

$$E(1) = \lim_{q \to 0} \frac{F_1}{q} = \frac{1}{4\pi\varepsilon_0}\sum_j \frac{q_i}{r_{1j}^2}\hat{\mathbf{r}}_{1j} \tag{2.4}$$

This vector equation is a shorthand version of three much longer

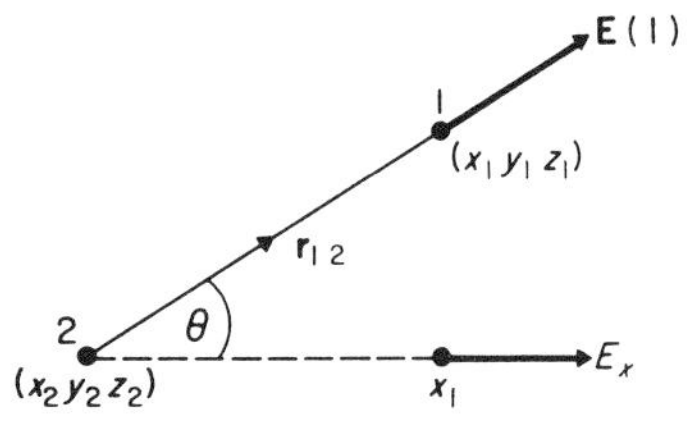

Fig. 2.2 The x-component E_x of electric field vector ***E***.

equations, which are nevertheless needed when a particular case has to be worked out.

For each coordinate plane there is a component of $\boldsymbol{E}(1)$ such as the x-component $E_x = E\cos\theta$ shown in Fig. 2.2. These components are, therefore, for a charge q_2 at $(x_2 y_2 z_2)$, given by

$$E_x(x_1 y_1 z_1) = \frac{q_2}{4\pi\varepsilon_0} \frac{(x_1 - x_2)}{\{(x_1 - x_2)^2 + (y_1 - y_2)^2 + (z_1 - z_2)^2\}^{3/2}},$$

$$E_y(x_1 y_1 z_1) = \frac{q_2}{4\pi\varepsilon_0} \frac{(y_1 - y_2)}{\{(x_1 - x_2)^2 + (y_1 - y_2)^2 + (z_1 - z_2)^2\}^{3/2}}$$

$$E_z(x_1 y_1 z_1) = \frac{q_2}{4\pi\varepsilon_0} \frac{(z_1 - z_2)}{\{(x_1 - x_2)^2 + (y_1 - y_2)^2 + (z_1 - z_2)^2\}^{3/2}}$$

Just writing out (2.4) in this way for Cartesian coordinates shows how useful vector equations are in saving time and space in print.

In a similar way we can write a vector equation for a charge distribution, using the notation shown in Fig. 2.3, where $\rho(x, y, z)$ is the charge density, which produces a charge $\rho\,\mathrm{d}\tau$ in a small volume $\mathrm{d}\tau$. From (2.4) the electric field at point 1 is now

$$\boldsymbol{E}(1) = \frac{1}{4\pi\varepsilon_0} \int_{\substack{\text{all}\\ \text{space}}} \frac{\rho(2)\mathrm{d}\tau_2}{r_{12}^2} \hat{\mathbf{r}}_{12} \tag{2.5}$$

where r_2 is the variable and the integral $\int \mathrm{d}\tau$ stands for $\iiint \mathrm{d}x\,\mathrm{d}y\,\mathrm{d}z$ in Cartesian coordinates and similar triple integrals for other coordinate systems. To apply this equation one must again evaluate each component, for example

$$E_y(x_1 y_1 z_1) = \frac{1}{4\pi\varepsilon_0} \int_{\substack{\text{all}\\ \text{space}}} \frac{(y_1 - y_2)\rho(x_2 y_2 z_2)\mathrm{d}x_2\,\mathrm{d}y_2\,\mathrm{d}z_2}{\{(x_1 - x_2)^2 + (y_1 - y_2)^2 + (z_1 - z_2)^2\}^{3/2}}$$

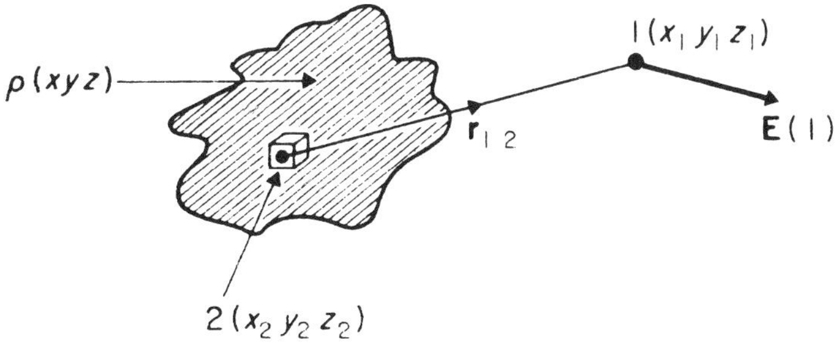

Fig. 2.3 Electric field $\boldsymbol{E}(1)$ at point 1 due to a distribution of charge.

Equations (2.4) and (2.5) show, in principle, how all electrostatic fields can be obtained. Until the charges move there is no more to electricity: it is just Coulomb's law and the principle of superposition. In practice there are some clever tricks to avoid such horrible calculations that are only fit for computers. Remember too that, whatever happens, electric charge is always conserved *in toto*, since it depends ultimately on the stability of the electrons and protons in the universe. (Recent theories of elementary particles and of cosmology imply that the proton is not absolutely stable, but has a half-life $\sim 10^{31}$ years.)

2.2 GAUSS'S LAW

Gauss's law is about electric flux. The idea of the flux of a vector field arises from the flow of a fluid. We need a measure of the field lines (lines of force) coming out of a surface. We know that if we tilt the surface it has a maximum 'flow' when it is normal to the field lines and minimum (zero) when it is parallel to them (Fig. 2.4). If we describe the size of the surface by $\mathrm{d}S$ and its orientation by its unit normal vector $\hat{\mathbf{n}}$, then the flux of the vector $\boldsymbol{E}$ from the element $\mathrm{d}S$ is defined as $\boldsymbol{E}\cdot\hat{\mathbf{n}}\,\mathrm{d}S$. This is commonly abbreviated to $\boldsymbol{E}\cdot\mathbf{d}\boldsymbol{S}$, where $\mathbf{d}\boldsymbol{S} = \hat{\mathbf{n}}\,\mathrm{d}S$ is a vector along the outward normal for outgoing flux. Therefore, for a point charge q at the origin using (2.4) we have

$$\boldsymbol{E}\cdot\mathbf{d}\boldsymbol{S} = \frac{q}{4\pi\varepsilon_0 r^2}\,\hat{\mathbf{r}}\cdot\mathbf{d}\boldsymbol{S} \tag{2.6}$$

The simplest way to evaluate this vector equation is to draw a sphere of radius r round the point charge and use spherical

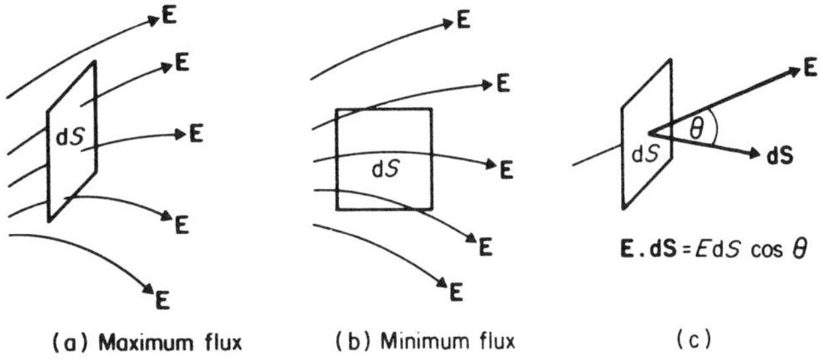

Fig. 2.4 Electric flux through various surfaces: (a) maximum for $\mathbf{d}\boldsymbol{S}$ normal to $\boldsymbol{E}$; (b) minimum for $\mathbf{d}\boldsymbol{S}$ in plane of $\boldsymbol{E}$; (c) flux of $\boldsymbol{E}$ is the scalar product $\boldsymbol{E}\cdot\mathbf{d}\boldsymbol{S}$.

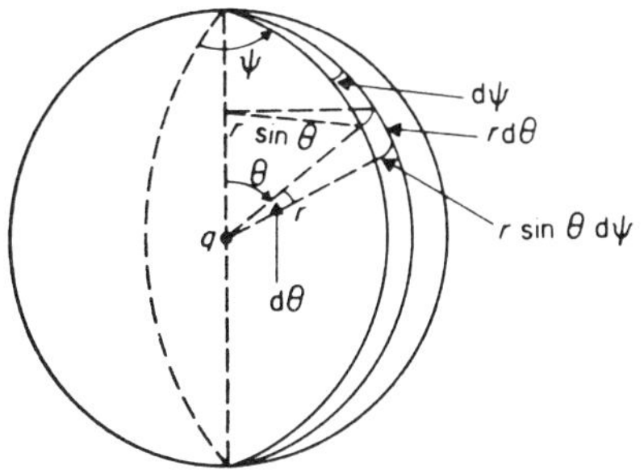

Fig. 2.5 Elementary surface area on a sphere around a point charge q.

polar coordinates (r, θ, ψ), as in Fig. 2.5. The elementary area $dS = r\sin\theta\, d\psi \cdot r\, d\theta$ and so the flux due to q through dS is

$$\frac{q}{4\pi\varepsilon_0 r^2}\hat{\mathbf{r}}\cdot\mathbf{dS} = \frac{q}{4\pi\varepsilon_0}\sin\theta\, d\psi\, d\theta$$

since $\hat{\mathbf{r}}$ and $\mathbf{dS}$ are parallel. Thus for an inverse square law of force, the flux from a point charge is independent of the distance r of the sphere from the point charge. The total flux through the sphere is

$$\int_S \boldsymbol{E}\cdot\mathbf{dS} = \int_S \frac{q}{4\pi\varepsilon_0}\sin\theta\, d\theta\, d\psi$$

where the integral is taken over the surface S of the sphere. This is easily evaluated:

$$\int_S \boldsymbol{E}\cdot\mathbf{dS} = \frac{q}{4\pi\varepsilon_0}\int_0^{\pi}\sin\theta\, d\theta\int_0^{2\pi} d\psi = \frac{q}{\varepsilon_0} \tag{2.7}$$

Of course a sphere is a particularly symmetrical surface to have chosen to find the total electric flux. Does it have to be so symmetrical to get such a simple answer? To find out we must evaluate (2.6) for

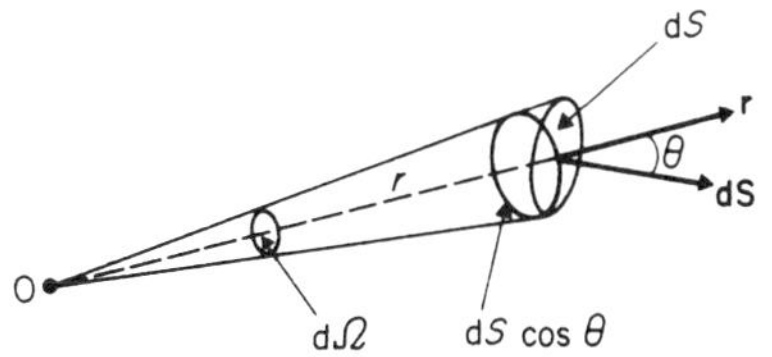

Fig. 2.6 Solid angle $d\Omega$ for cone of base dS.

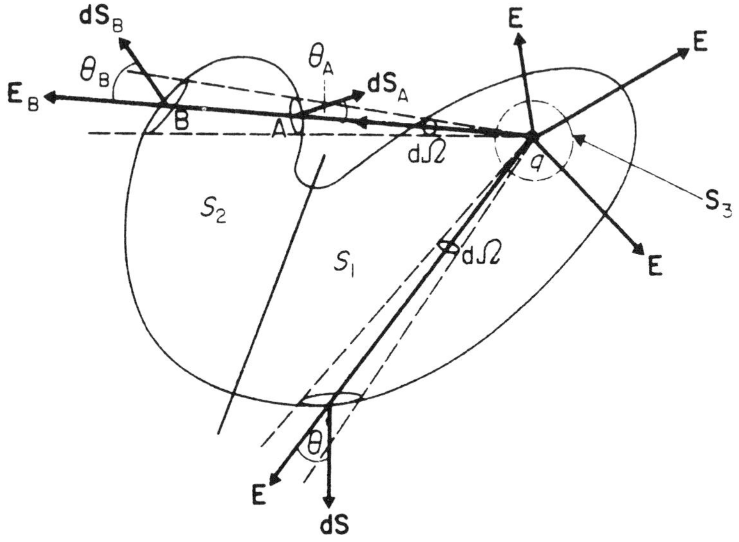

Fig. 2.7 Arbitrary surface around a point charge q.

a surface of arbitrary shape (Fig. 2.7) and this is best done by using the concept of a solid angle $d\Omega$ (Fig. 2.6). For any surface area dS whose normal $\mathbf{dS}$ makes an angle θ with the radius vector $\boldsymbol{r}$ from an arbitrary point O, the solid angle $d\Omega$ is $d\Omega = dS\cos\theta/r^2$. We can therefore write (2.6) as

$$\boldsymbol{E}\cdot\mathbf{dS} = q\,d\Omega/4\pi\varepsilon_0$$

and use this expression for the electric flux through dS to evaluate the total flux through the arbitrary surface S.

We first divide S into two parts: S_1 that does enclose q; and S_2 that doesn't. Then for S_1 we must compute the $\int_{S_1} d\Omega$. Since q is a point charge, any surface S_1 that completely encloses q will subtend the same solid angle at q as the sphere S_3. Therefore,

$$\text{Flux through } S_1 = \frac{q}{4\pi\varepsilon_0}\int_{S_3} d\Omega = \frac{q}{\varepsilon_0}$$

For the surface S_2 that does not enclose q, any flux cone must cut the surface twice, once on entry (e.g. at A) and once on exit (e.g. at B). The total flux flowing *out* of the surface bounded by A, B and the cone in between is therefore $\boldsymbol{E}_{\mathrm{B}}\cdot\mathbf{d}\boldsymbol{S}_{\mathrm{B}} - \boldsymbol{E}_{\mathrm{A}}\cdot\mathbf{d}\boldsymbol{S}_{\mathrm{A}}$. But

$$\frac{E_{\mathrm{B}}}{E_{\mathrm{A}}} = \frac{r_{\mathrm{A}}^2}{r_{\mathrm{B}}^2}$$

by the inverse square law for electric fields and

$$\frac{\mathrm{d}S_{\mathrm{B}}\cos\theta_{\mathrm{B}}}{\mathrm{d}S_{\mathrm{A}}\cos\theta_{\mathrm{A}}} = \frac{r_{\mathrm{B}}^2\,\mathrm{d}\Omega}{r_{\mathrm{A}}^2\,\mathrm{d}\Omega}$$

by definition of solid angles. The flux $\boldsymbol{E}_{\mathrm{A}}\cdot\mathbf{d}\boldsymbol{S}_{\mathrm{A}}$ that flows into this region is thus exactly the same as the flux $\boldsymbol{E}_{\mathrm{B}}\cdot\mathbf{d}\boldsymbol{S}_{\mathrm{B}}$ that flows out of the region and the net flux for the surface S_2:

$$\int_{S_2} \boldsymbol{E}\cdot\mathbf{d}\boldsymbol{S} = 0$$

It follows, then, that the total flux for *any* surface surrounding q is

$$\int_{S} \boldsymbol{E}\cdot\mathbf{d}\boldsymbol{S} = \frac{q}{\varepsilon_0} \tag{2.8}$$

exactly as obtained from the sphere in (2.7).

By the principle of superposition the flux due to two charges q_1 and q_2 is just

$$\int_{S} \boldsymbol{E}_1\cdot\mathbf{d}\boldsymbol{S} + \int_{S} \boldsymbol{E}_2\cdot\mathbf{d}\boldsymbol{S} = \frac{1}{\varepsilon_0}(q_1 + q_2)$$

and it follows that the flux due to any charge distribution is

$$\int_{S} \boldsymbol{E}\cdot\mathbf{d}\boldsymbol{S} = \frac{1}{\varepsilon_0}\int_{V} \rho\,\mathrm{d}\tau \tag{2.9}$$

where V is the volume enclosed by S.

This is *Gauss's law:* the total flux out of any closed surface is equal to the total charge enclosed by it divided by the electric constant, ε_0.

2.2.1 Applications of Gauss's Law

Gauss's law is particularly useful for finding the electric field due to a symmetrical distribution of charge. In each case the Gaussian surface is chosen to suit the symmetry of the problem, as will be seen from three examples.

1. ***E** for a sphere of charge*

Suppose we have a sphere of uniform charge density ρ_0, then

$$Q = \int_\tau \rho \, d\tau = \frac{4}{3}\pi a^3 \rho_0$$

as illustrated in Fig. 2.8, is its total charge.

We first draw an imaginary Gaussian surface S of radius R through the point P, where we wish to find $\boldsymbol{E}$. Since the charge is uniformly distributed throughout Q, by symmetry $\boldsymbol{E}$ is everywhere radial from Q.

Applying Gauss's law we obtain

$$\int_S \boldsymbol{E}\cdot\mathbf{d}\boldsymbol{S} = \frac{Q}{\varepsilon_0}$$

Hence, $$E\cdot 4\pi R^2 = Q/\varepsilon_0$$

or $$E = Q/4\pi\varepsilon_0 R^2$$

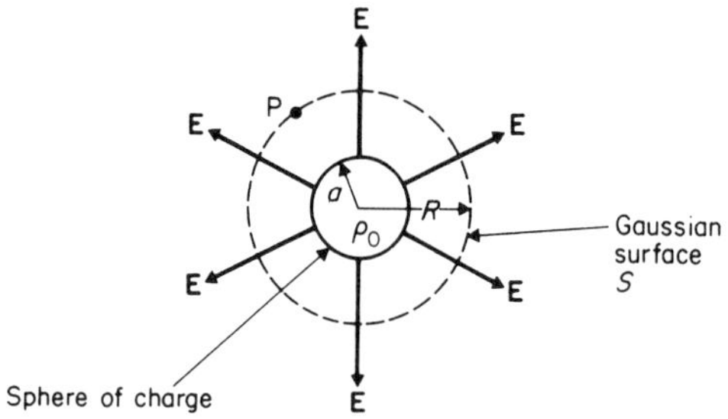

Fig. 2.8 Electric field of a sphere of charge.

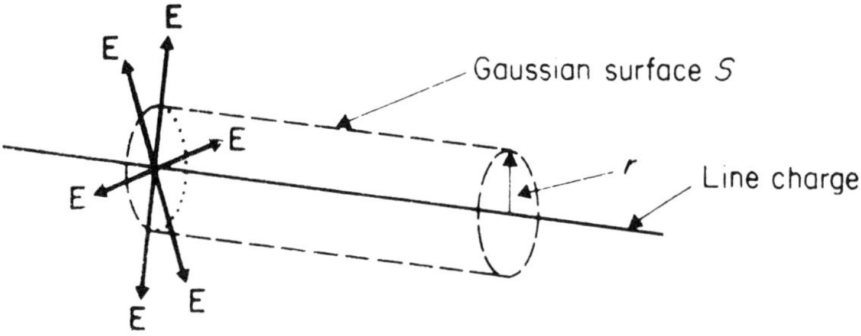

Fig. 2.9 Electric field from a line of charge.

This is exactly the same as the field due to a point charge Q at the centre of the sphere of charge, a result that is quite hard to prove without Gauss's law.

2. ***E** for a line charge*
To find the electric field at a point distance r from an infinite line charge $\lambda\,\mathrm{C\,m^{-1}}$, we note the cylindrical symmetry of the problem and draw a Gaussian cylindrical surface S (Fig. 2.9) of radius r and of length 1 m. The electric vectors will be everywhere radial by symmetry and the same at all points along the line. Applying Gauss's law,

$$\int_S \boldsymbol{E}\cdot\mathbf{d}\boldsymbol{S} = \int_{\substack{\text{cylindrical}\\\text{surface}}} \boldsymbol{E}\cdot\mathbf{d}\boldsymbol{S} + \int_{\substack{\text{end}\\\text{faces}}} \boldsymbol{E}\cdot\mathbf{d}\boldsymbol{S} = \frac{\lambda}{\varepsilon_0}$$

Hence, $$2\pi r E + 0 = \lambda/\varepsilon_0$$

or $$E = \lambda/2\pi\varepsilon_0 r$$

3. ***E** for a plane sheet of charge*
To find the electric field near a plane sheet of charge $\sigma\,\mathrm{C\,m^{-2}}$, we first note that $\boldsymbol{E}$ must be everywhere normal to the sheet and that the field $\boldsymbol{E}_1$ on one side must be the same size as the field $\boldsymbol{E}_2$ on the other side (Fig. 2.10). By symmetry our Gaussian surface S is a rectangular box whose sides parallel to the sheet of area A contain all the flux that the charges are producing. By Gauss's law,

$$\int_S \boldsymbol{E}\cdot\mathbf{d}\boldsymbol{S} = \int_A \boldsymbol{E}_1\cdot\mathbf{d}\boldsymbol{S} + \int_A \boldsymbol{E}_2\cdot\mathbf{d}\boldsymbol{S} = \frac{\sigma A}{\varepsilon_0}$$

Here both these integrals refer to outward fluxes, so

$$EA + EA = \sigma A/\varepsilon_0$$

or $$E = \sigma/2\varepsilon_0$$

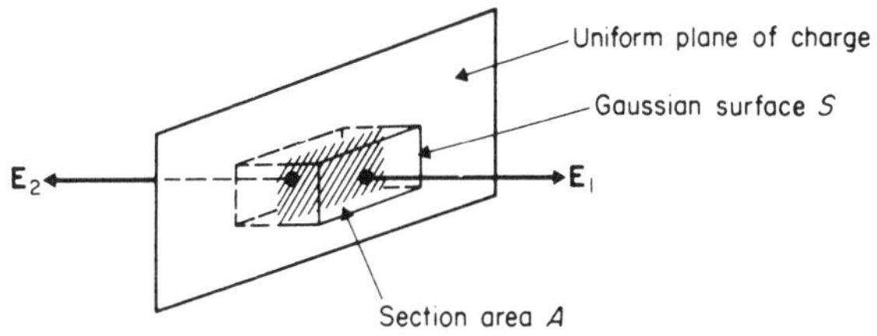

Fig. 2.10 Electric field from a sheet of charge.

2.3 ELECTRIC POTENTIAL

The electrostatic field, like the gravitational field, is a conservative field. In other words it is a field of force in which the work done in taking a particle from one point to another is independent of the path taken between them. The concept of potential energy used in gravitational problems can therefore be applied to electrical problems. In mechanics the work done, dW, on a particle travelling a distance ds along a path ab by an applied force $\boldsymbol{F}$ (Fig. 2.11) is given by the component of $\boldsymbol{F}$ acting in the direction s times ds,

$$dW = F\cos\theta\, ds$$

The total work done over the path ab is then

$$W = \int_a^b (F\cos\theta)ds = \int_a^b \boldsymbol{F}\cdot\mathbf{d}\boldsymbol{s}$$

where $\mathbf{d}\boldsymbol{s}$ is a vector element of the line from a to b. In electrostatics the particle becomes a test charge moving quasi-statically (with zero velocity) from a to b and the applied force must overcome the electric forces acting on the test charge. In Fig. 2.12 the electric force on the test charge at P is $\boldsymbol{E}_{\mathrm{p}}$ and so the external, applied force to move the charge quasi-statically is $-\boldsymbol{E}_{\mathrm{p}}$. It is this force that is needed to calculate the external work done against the electric field due to the point charge q.

Therefore the work done on unit charge in taking it from a to b is

$$\int_a^b -\boldsymbol{E}\cdot\mathbf{d}\boldsymbol{s}$$

Using the definition of $\boldsymbol{E}$ in (2.4) and noting that $\hat{\mathbf{r}}\cdot\mathbf{d}\boldsymbol{s} = dr$, this integral becomes

$$\frac{-q}{4\pi\varepsilon_0}\int_a^b \frac{1}{r^2}dr = \frac{-q}{4\pi\varepsilon_0}\left(\frac{1}{r_{\mathrm{a}}} - \frac{1}{r_{\mathrm{b}}}\right)$$

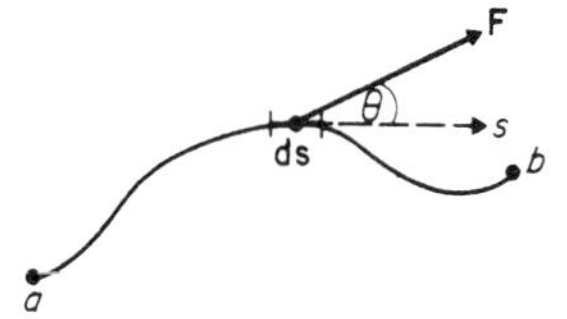

Fig. 2.11 Force $\boldsymbol{F}$ acting on a particle as it moves along the path ab.

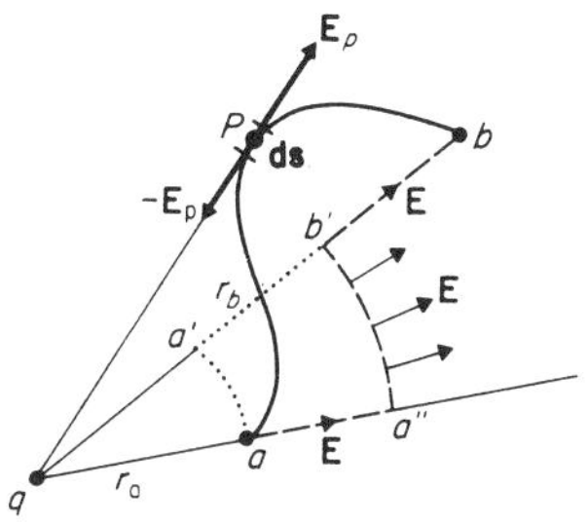

Fig. 2.12 Electric force $\boldsymbol{E}_P$ and external force $-\boldsymbol{E}_P$ acting on a test charge at P, used in calculating the work done on taking unit charge from a to b.

Referring to Fig. 2.12 we can see that this amount of work would also be done if the charge was just moved radially from a' to b. Equally it would be the same if it was moved first along the radial path aa'', then along the circular path $a''b'$ and finally along the radial path $b'b$, since $\boldsymbol{E}$ is always normal to a circular path about q and therefore no work is done along a circular path.

It thus follows that the $\int_a^b \boldsymbol{E}\cdot\mathbf{d}\boldsymbol{s}$ is the same along any arbitrary path which can always be considered as the zero sum of normal components along circular paths and the work done by the tangential components along radial paths. If the path is a closed loop C (a to b and back to a) then clearly the integral is zero:

$$\oint_C \boldsymbol{E}\cdot\mathbf{d}\boldsymbol{s} = 0 \tag{2.10}$$

This is the *circulation law* for the electrostatic field and is a characteristic of conservative fields that have spherical symmetry and of forces that are radial $f(\boldsymbol{r})$. It does not have to be an inverse square law force to have zero circulation.

When the path is not closed, the work done depends only on the end points and is independent of the path taken (Fig. 2.12). The work done on unit charge can therefore be represented as the difference between two electric potentials $\phi(b)$ and $\phi(a)$, by analogy with mechanical potential energy,

$$-\int_a^b \boldsymbol{E}\cdot\mathbf{d}\boldsymbol{s} = \phi(b) - \phi(a) \tag{2.11}$$

To obtain an absolute value for the electric potential, we must specify its zero. This is taken for convenience to be at an infinite distance from the source q so that

$$-\int_{\infty}^{r} \boldsymbol{E}\cdot \mathbf{d}\boldsymbol{s} = \phi(r) \tag{2.12}$$

defines the potential $\phi(r)$ at any point of distance r from q. Using (2.4) for $\boldsymbol{E}$, we obtain

$$\phi(r) = \frac{-q}{4\pi\varepsilon_0}\left(\frac{1}{r}\right)$$

for a point charge q at the origin.

Electric potentials can be superimposed like electric fields and so for a distribution of charges we obtain similar to (2.4) and (2.5), namely

$$\phi(1) = \frac{1}{4\pi\varepsilon_0}\sum_i \frac{q_i}{r_{1i}} \tag{2.13}$$

and

$$\phi(1) = \frac{1}{4\pi\varepsilon_0}\int_{\substack{\text{all}\\ \text{space}}} \frac{\rho(2)\mathrm{d}\tau_2}{r_{12}} \tag{2.14}$$

It is important to remember that electric potentials are the work done on unit charges and therefore measured in volts, not joules like potential energy. The volt is defined by: the work done is 1 joule when a charge of 1 coulomb is moved through a potential difference of 1 volt.

The calculation of electric fields can often be achieved more simply from electric potentials than from (2.4) and (2.5). To do this we must invert (2.11) to obtain a differential equation. For two points distance Δx apart, by the definition of electric potential, the work done on moving charge through Δx is

$$\begin{aligned}\Delta W &= \phi(x+\Delta x, y, z) - \phi(xyz)\\ &= \left(\frac{\partial \phi}{\partial x}\right)_{y,z} \Delta x\end{aligned}$$

But the work done against the electric field $\boldsymbol{E}$ is

$$\Delta W = -\int_{x}^{x+\Delta x} \boldsymbol{E}\cdot \mathbf{d}\boldsymbol{s} = -E_x \Delta x$$

Hence

$$E_x = -\frac{\partial \phi}{\partial x}$$

Similarly, for movements along Δy and Δz we find

$$E_y = -\frac{\partial \phi}{\partial y} \quad \text{and} \quad E_z = -\frac{\partial \phi}{\partial z}$$

so that

$$\boldsymbol{E} = -\left(\frac{\partial \phi}{\partial x}\hat{\mathbf{i}} + \frac{\partial \phi}{\partial y}\hat{\mathbf{j}} + \frac{\partial \phi}{\partial z}\hat{\mathbf{k}}\right) \tag{2.15}$$

where $(\hat{\mathbf{i}}, \hat{\mathbf{j}}, \hat{\mathbf{k}})$ are unit vectors along the Cartesian axes $0x, 0y, 0z$.

In vector calculus the gradient of a scalar field $\Omega(xyz)$ that is continuously differentiable is defined as the vector

$$\text{grad}\, \Omega = \frac{\partial \Omega}{\partial x}\hat{\mathbf{i}} + \frac{\partial \Omega}{\partial y}\hat{\mathbf{j}} + \frac{\partial \Omega}{\partial z}\hat{\mathbf{k}} \tag{2.16}$$

Comparing (2.15) and (2.16) we see that the electrostatic field $\boldsymbol{E}$ is just minus the gradient of the electric potential ϕ:

$$\boldsymbol{E} = -\,\text{grad}\, \phi \tag{2.17}$$

It follows that electric fields have the convenient unit of volts per metre, as well as the more fundamental one of newtons per coulomb from (2.4).

2.3.1 Conductors

A metal may be considered as a conductor containing many 'free' electrons which can move about inside but not easily escape from the surface. Inside a metal there is perfect charge balance between the positive ions and the negative electrons and, on a macroscopic scale, the net charge density is zero. By Gauss's law the electric field $\boldsymbol{E}_i$ inside a Gaussian surface that coincides with the surface of a metallic conductor (Fig. 2.13(a)) must therefore also be zero.

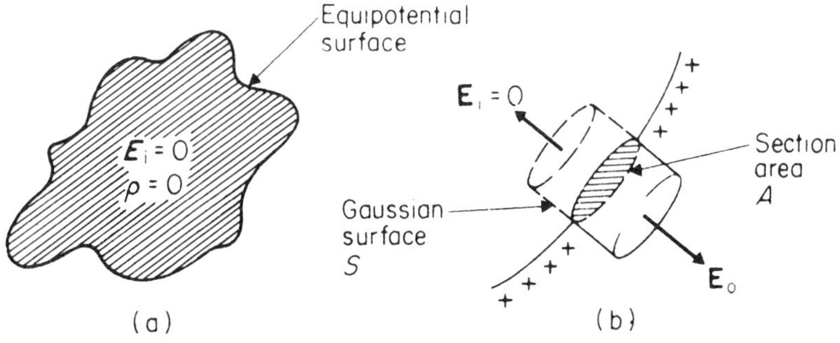

Fig. 2.13 (a) Electric field inside a conductor is zero. (b) Outside a charged conductor it depends on the charge density at the surface $\sigma\,\text{C m}^{-2}$.

By (2.17) the gradient of the electric potential at the surface, grad ϕ, must be zero and so the surface of a conductor is an equipotential surface and the interior of the conductor is an equipotential region ($\phi = \text{constant}$).

When a conductor is charged the excess charges stay on the surface, where they are not completely free. They then produce an external field $\boldsymbol{E}_0$ just outside the surface (Fig. 2.13(b)), whose value can be obtained from the cylindrical Gaussian surface S of cross-sectional area A and cylindrical axis normal to the surface. Clearly

$$\int_S \boldsymbol{E}\cdot\mathbf{d}\boldsymbol{S} = E_0 A + 0 = \frac{\sigma A}{\varepsilon_0}$$

and
$$E_0 = \sigma/\varepsilon_0 \tag{2.18}$$

This is just *twice* the field for a sheet of charge (section 2.2) because the internal electrons have produced zero internal field when the 'sheet of charge' is no longer isolated, but on a conductor. A similar result is seen if we consider two uniformly charged, parallel, conducting plates. In Fig. 2.14 the large Gaussian surface encloses a total charge of zero and so the external field $\boldsymbol{E}_0 = 0$. On the other hand the internal field $\boldsymbol{E}_i$, whether obtained from the positively charged plate, or the negatively charged plate, is just $E_i = \sigma/\varepsilon_0$.

An interesting question is: can the inner surface of a hollow conductor be charged? In Fig. 2.15 if there is a surface density of

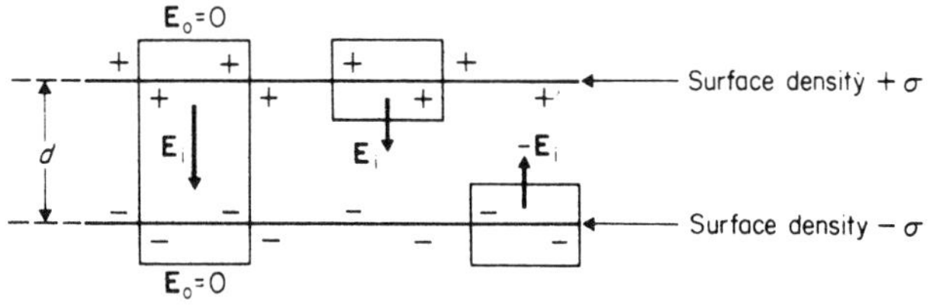

Fig. 2.14 Electric field between two charged plates.

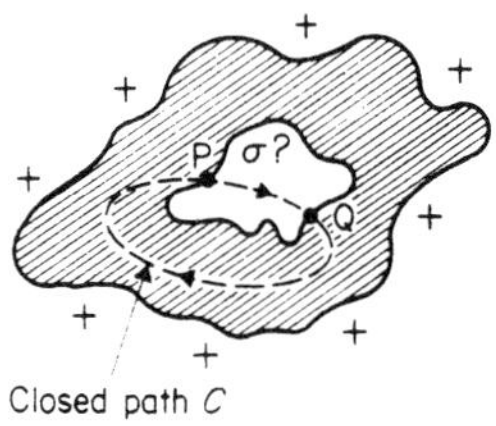

Fig. 2.15 Can the inner surface of a hollow conductor be charged?

charge σ inside the cavity then there is an electric field in the cavity and for the closed path C through P and Q:

$$\oint_C \boldsymbol{E}\cdot\mathbf{ds} \neq 0$$

But this would violate the circulation law (2.10) and so $\boldsymbol{E}$ inside a cavity must be zero and it is impossible to charge the inside of a hollow conductor. This important principle is the basis of electrostatic screening (the Faraday cage).

2.4 ELECTROSTATIC ENERGY

Electric energy is stored in capacitors, for example in parallel plate capacitors (Fig. 2.14). From (2.11) we see that the potential difference V between the plates, for a uniform field $\boldsymbol{E}_i$ between the plates distance $\boldsymbol{d}$ apart is

$$\boldsymbol{E}_i\cdot\boldsymbol{d} = \phi_+ - \phi_- = V$$

But $E_i = \sigma/\varepsilon_0$ and for a uniform distribution of charge, $\sigma = Q/A$, where Q is the total charge on the plates of area A.

Hence $$V = \left(\frac{d}{\varepsilon_0 A}\right)Q$$

that is $$V \propto Q$$

The proportionality between V and Q is always found for two oppositely charged conductors, since it arises from the principle of superposition: doubling the charges doubles the field and doubles the work done on taking unit charge from one plate to the other. By convention, we define *capacitance* by

$$Q = CV$$

and so for the parallel plate capacitor the capacitance is

$$C = \varepsilon_0 A/d \tag{2.19}$$

The unit of capacitance is the farad (named after Faraday) and is a coulomb per volt. It is a very large unit and typical small capacitors are in microfarad (μF) for use at low frequencies and in picofarad (pF) for radio frequencies. When capacitors are in parallel they are all at the same potential (Fig. 2.16(a)) and so $C = \sum_i C_i$, but when they are in series they each carry the same charge (Fig. 2.16(b)) and then $1/C = \sum_i 1/C_i$.

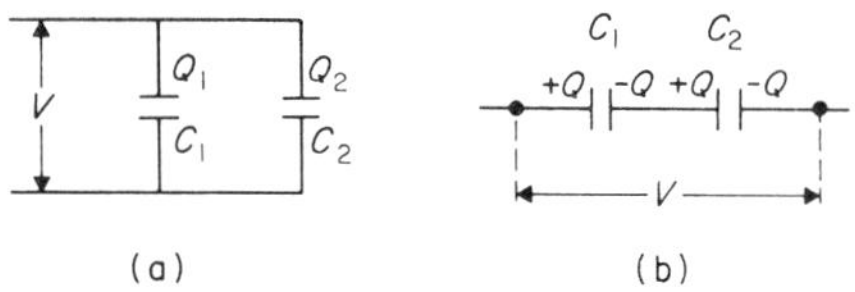

Fig. 2.16 Capacitors connected (a) in parallel, and (b) in series.

The work done in charging a capacitor is equal to the *energy*, U, stored in it and so

$$U = \int_0^Q V\,dQ = \frac{1}{C}\int_0^Q Q\,dQ = \frac{Q^2}{2C} = \frac{1}{2}CV^2 \tag{2.20}$$

For the parallel plate capacitor, neglecting end-effects, the energy is:

$$U = \frac{Q^2}{2C} = \frac{(\sigma A)^2}{2(\varepsilon_0 A/d)} = \frac{\sigma^2 Ad}{2\varepsilon_0}$$

The *energy density*, u, in this electrostatic field is the total energy U divided by the volume Ad and so

$$u = \frac{\sigma^2}{2\varepsilon_0} = \frac{1}{2}\varepsilon_0 E^2 \tag{2.21}$$

since $E = \sigma/\varepsilon_0$ between the plates. This expression for the energy density was derived from a specific example, the parallel plate capacitor, but it contains only the electric field E and the electric constant ε_0. It is, in fact, quite general, as can be seen from Fig. 2.17. There an electric field is described by a number of equipotentials $\phi_1, \phi_2, \ldots$ and we can imagine any small volume $d\tau$ in that field as

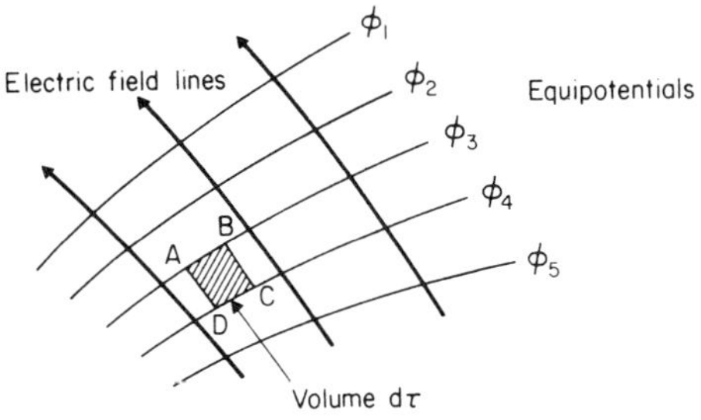

Fig. 2.17 Equipotentials in an electric field.

a tiny parallel plate capacitor ABCD, since a conducting plate is an equipotential surface. Therefore quite generally

$$U = \frac{1}{2}\varepsilon_0 \int_\tau E^2 \, d\tau \qquad (2.22)$$

The idea that the electrostatic energy is stored *in the electric field* is an important one and enables the energy to be computed without knowing anything about the distribution of electric charge. It is even more important in discussing the energy of radio waves. Clearly radio stations transmit electromagnetic energy in the waves we receive at our aerials, but they do not transmit electric charges over long distances. The energy is stored, and travels, in the electromagnetic field of the wave.

There is one point of difficulty in calculating electric fields. The energy of a charged sphere of radius a (problem 2) is $\frac{1}{2}Q^2/(4\pi\varepsilon_0 a)$ and so the self-energy of a point charge ($a \to 0$) would be infinite! Obviously the idea of stored energy in an electrostatic field is not consistent with the presence of point charges: either the electron has a finite size or we cannot extend our concept to elementary charged particles. To avoid this real difficulty we compute the energy only of the electrostatic fields between the charges, and omit their self-energy. Thus we can write (2.22) more generally as

$$U = \frac{1}{2}\varepsilon_0 \int_\tau \mathbf{E}\cdot\mathbf{E} \, d\tau \qquad (2.23)$$

2.4.1 Electric stress

The limit to the energy that can be stored in a particular capacitor depends on the maximum electric field E that its insulation will withstand before it breaks down under the electric stress. Typically the insulation strength is about $10^8\ \mathrm{V\,m^{-1}}$ so that a large capacitor of internal volume $0.1\ \mathrm{m^3}$ would hold a maximum of $\frac{1}{2}\varepsilon_0 \times 10^{16} \times 0.1 \simeq 5\ \mathrm{kJ}$. This is small compared with the 10 MJ of chemical energy in 1 kg of common salt and very small compared with the 50 TJ of nuclear energy from the fission of 1 kg of uranium.

Still larger capacitors cannot easily be built because of the enormous mechanical stresses they are subjected to when charged. This can be seen by applying the principle of virtual work to charged capacitor plates (Fig. 2.18). If $\mathbf{F}$ is the attractive force between the plates, then the external work done ΔU increasing the separation by Δx must

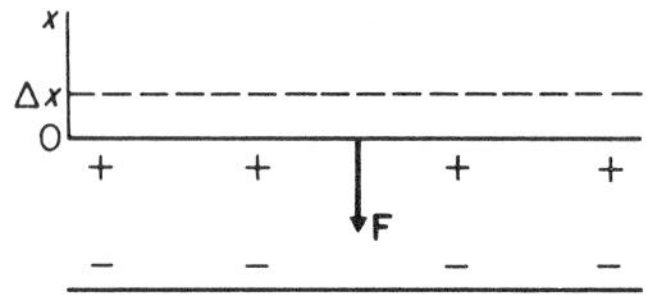

Fig. 2.18 Work is done in separating charged capacitor plates.

equal the change in electrostatic energy for constant Q. Using (2.19) and (2.20),

$$\Delta U = -\boldsymbol{F}\cdot\Delta \boldsymbol{x} = \frac{1}{2}Q^2\Delta\left(\frac{1}{C}\right) = \frac{Q^2\Delta x}{2\varepsilon_0 A}$$

or

$$\boldsymbol{F} = -(Q^2/2\varepsilon_0 A)\hat{x}$$

But the charge $Q = \sigma A$ and the electric field is $E = \sigma/\varepsilon_0$, so that the magnitude of the stress is

$$\frac{F}{A} = \frac{1}{2}\varepsilon_0 E^2$$

This is the same as the energy density (2.21) and so for $E = 10^8\,\mathrm{V\,m^{-1}}$, the mechanical stress is $\simeq 50\,\mathrm{kN\,m^{-2}}$ or about 5 tonne weight per square metre.

2.5 DIELECTRICS

Dielectric materials – like glass, paper and plastics – are electrical insulators. They have no free charges and do not conduct electricity, but they do influence electric fields. Faraday discovered that inserting an insulator between the plates of a parallel plate capacitor increased its capacitance. We define the *relative permittivity* (dielectric constant), ε_r, of an insulator from Faraday's experiment as the ratio of the capacitances when the capacitor is completely filled by the insulator to when it is empty. Thus,

$$\varepsilon_r = C_{\text{full}}/C_{\text{empty}} \tag{2.24}$$

For commercial dielectrics the material is normally of uniform composition and when used in low electric fields is a linear, isotropic, homogeneous medium characterized by a single constant ε_r. Typical values of ε_r are air = 1.0006, polythene = 2.3, glass = 6, barium titanate ceramic = 3000. In designing capacitors, transformers, coaxial cables, etc., ε_r is an important factor influencing the design.

Physically there is a great deal to investigate in the dielectric behaviour of gases, liquids and, especially, solids, where ε_r can be an anisotropic parameter described by an appropriate tensor. The permittivity will also vary with temperature, with the frequency of an electromagnetic field and will become nonlinear in high electric fields. It is only in its familiar usage in low-frequency, low-field capacitors that it can be treated as a dielectric constant.

2.5.1 Polarization

What happens when a slab of dielectric is inserted into a parallel plate capacitor (Fig. 2.19)? We know that the capacitance $C = Q/V$ increases from Faraday's experiment and so, if the charge Q on the plates has not leaked away, the potential V must have been reduced. Ignoring end-effects there is a uniform electric field $E = V/d$, so the electric field in the presence of the dielectric E_d must be less than that originally present, E_0. We can explain this by postulating an induced surface charge density, σ_p, on each side of the dielectric slab, providing the charges are of opposite sign to those inducing them from the respective capacitor plates. If these free charges have surface density σ_f, then

$$E_d = \frac{\sigma_f - \sigma_p}{\varepsilon_0} < \frac{\sigma_f}{\varepsilon_0} = E_0 \tag{2.25}$$

using the expression $E = \sigma/\varepsilon_0$, which depends only on Gauss's law. This process is called *polarization* of the dielectric and occurs only in the presence of the electric field (the part of the dielectric outside the capacitor plates is not polarized, Fig. 2.19).

One way in which a neutral atom can acquire a dipole moment is shown in Fig. 2.20. The spherical electron cloud of the neutral atom is distorted by the applied electric field $\boldsymbol{E}$ and this distorted

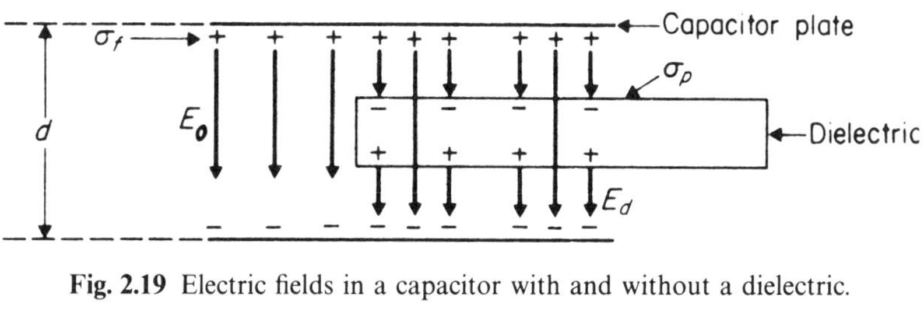

Fig. 2.19 Electric fields in a capacitor with and without a dielectric.

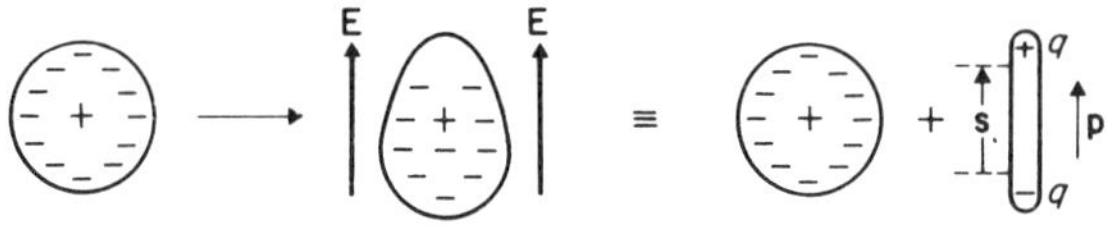

Fig. 2.20 Polarization of an atom gives it a dipole moment $\boldsymbol{p}$.

charge distribution is equivalent, by the principle of superposition, to the original spherical distribution plus a dipole distribution whose *dipole moment* is

$$\boldsymbol{p} = q\boldsymbol{s} \tag{2.26}$$

where $\boldsymbol{s}$ is the distance vector from $-q$ to $+q$ of the dipole.

If a polarized dielectric consists of N such dipoles per unit volume, then we define its *polarization* $\boldsymbol{P}$ as

$$\boldsymbol{P} = N\boldsymbol{p} = \sum_i \boldsymbol{p}_i/\tau$$

where τ is the total volume. The SI unit of dipole moment is the coulomb-metre and that of polarization coulomb per square metre. Going back to Fig. 2.19, we see that the surface charge on the dielectric is σ_p coulombs per square metre. If we assume this is due to N electrons per unit volume being displaced upwards a distance δ at each surface of area A', then the total charge is

$$\sigma_p A' = Ne\delta A'$$

or

$$\sigma_p = Ne\delta$$

But $Ne\delta$ is just Np and so $P = \sigma_p$ and (2.25) becomes

$$E_d = \frac{\sigma_f - P}{\varepsilon_0}$$

From this equation it is clear that $\varepsilon_0 E_d$ and P have the same dimensions. We define *electric susceptibility*, χ_e, a dimensionless quantity, as the ratio

$$\chi_e = P/\varepsilon_0 E_d$$

Knowing χ_e for our dielectric material we therefore obtain the reduced electric field E_d as

$$E_d = \frac{\sigma_f}{\varepsilon_0}\left(\frac{1}{1+\chi_e}\right) \tag{2.27}$$

The capacitance of the capacitor is inversely proportional to the potential and so to the electric field. Using (2.24), (2.25) and (2.27) we therefore obtain

$$\varepsilon_r = \frac{E_0}{E_d} = (1 + \chi_e) \tag{2.28}$$

Measurement of the electric susceptibility χ_e of matter at low frequencies is thus a measurement of the relative permittivity, ε_r, which at optical frequencies can be shown to be the square of the refractive index, as explained in Chapter 10.

2.5.2 Electric displacement

For all electrostatic systems we have the fundamental equation (2.9), Gauss's law:

$$\int_S \boldsymbol{E}\cdot\mathbf{d}\boldsymbol{S} = \frac{1}{\varepsilon_0}\int_V \rho\,\mathrm{d}\tau$$

where V is the volume enclosed by the surface S.

When dielectrics are present, the charge density ρ will be the sum of any polarization charges of density ρ_p and any free charges of density ρ_f. Therefore,

$$\varepsilon_0 \int_S \boldsymbol{E}\cdot\mathbf{d}\boldsymbol{S} = \int_V \rho_f\,\mathrm{d}\tau + \int_V \rho_p\,\mathrm{d}\tau \tag{2.29}$$

We have seen (Fig. 2.19) that in a parallel plate capacitor, when the polarization $\boldsymbol{P}$ is normal to the surface of the dielectric, its magnitude is just the surface density of charge σ_p displaced from

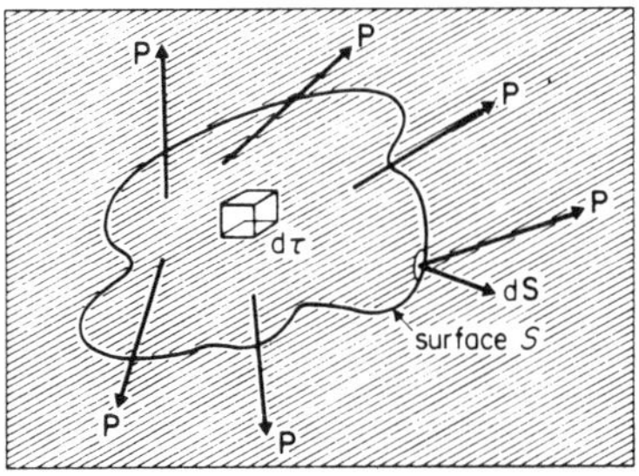

Fig. 2.21 Non-uniform polarization of a dielectric.

inside the dielectric. At the ends of the dielectric slab where $\boldsymbol{P}$ is tangential to the surface the surface density of charge is zero. It is the normal component of $\boldsymbol{P}$ that produces a surface charge, so that for an arbitrary surface S inside a dielectric (Fig. 2.21) the charge dq_p displaced across a surface element $\mathbf{d}\boldsymbol{S}$ is $\boldsymbol{P}\cdot\mathbf{d}\boldsymbol{S}$. A nonuniform polarization at the surface S therefore produces a total displacement of charge q_p across S given by

$$q_p = \int_S \boldsymbol{P}\cdot\mathbf{d}\boldsymbol{S}$$

Since the dielectric is electrically neutral this is compensated by a volume density of charge ρ_p such that

$$\int_V \rho_p \, d\tau = -q_p$$

Hence, the flux of $\boldsymbol{P}$ is given by a type of Gauss's law for polarized dielectrics:

$$\int_S \boldsymbol{P}\cdot\mathbf{d}\boldsymbol{S} = -\int_V \rho_p \, d\tau$$

Combining this with (2.29), we have

$$\int_S (\varepsilon_0 \boldsymbol{E} + \boldsymbol{P})\cdot\mathbf{d}\boldsymbol{S} = \int_V \rho_f \, d\tau$$

and define the *electric displacement*, $\boldsymbol{D}$, as

$$\boldsymbol{D} = \varepsilon_0 \boldsymbol{E} + \boldsymbol{P} = \varepsilon_0(1 + \chi_e)\boldsymbol{E} \qquad (2.30)$$

so that Gauss's law can also be written

$$\int_S \boldsymbol{D}\cdot\mathbf{d}\boldsymbol{S} = \int_V \rho_f \, d\tau \qquad (2.31)$$

The flux of $\boldsymbol{D}$ thus depends solely on the free charges and this can be very useful, for example, in microwave physics.

However if we use this equation and, from (2.28) and (2.30), write

$$\boldsymbol{D} = \varepsilon_r \varepsilon_0 \boldsymbol{E} \qquad (2.32)$$

then we must remember that for many materials ε_r is not just a number. As we have emphasized before, $\boldsymbol{P}$ (and hence $\boldsymbol{D}$) is not proportional to $\boldsymbol{E}$ for nonlinear materials and, in any case, χ_e (and so ε_r) can vary with frequency, temperature, crystal direction, etc.

This is one reason why we refer to ε_r as the relative permittivity rather than the dielectric constant of a dielectric.

2.5.3 Boundary relations

What happens to the electric field when it crosses the boundary between two dielectrics of permittivities $\varepsilon_1 = \varepsilon_{r1}\varepsilon_0$ and $\varepsilon_2 = \varepsilon_{r2}\varepsilon_0$? To find out we apply Gauss's law as given in (2.31) and the circulation law, (2.10), to the electric vectors $\boldsymbol{D}$ and $\boldsymbol{E}$ shown in Fig. 2.22.

For the flux into and out of the Gaussian cylinder of cross-section dS and negligible height we have

$$\boldsymbol{D}_1 \cdot \mathbf{d}\boldsymbol{S}_1 + \boldsymbol{D}_2 \cdot \mathbf{d}\boldsymbol{S}_2 = 0$$

since there are no free charges in dielectrics. Hence only the normal component D_n of each electric displacement contributes and

$$D_{1n} = D_{2n} \tag{2.33}$$

Applying the circulation law to the electric fields $\boldsymbol{E}_1$ and $\boldsymbol{E}_2$ crossing the closed loop of length $s_1 + s_2$, we have

$$\oint_C \boldsymbol{E} \cdot \mathbf{d}\boldsymbol{s} = \boldsymbol{E}_1 \cdot \boldsymbol{s}_1 + \boldsymbol{E}_2 \cdot \boldsymbol{s}_2 = 0$$

Since we can contract the loop to be as near as we wish to the surface, only the tangential component E_t of each electric field contributes and

$$E_{1t} = E_{2t} \tag{2.34}$$

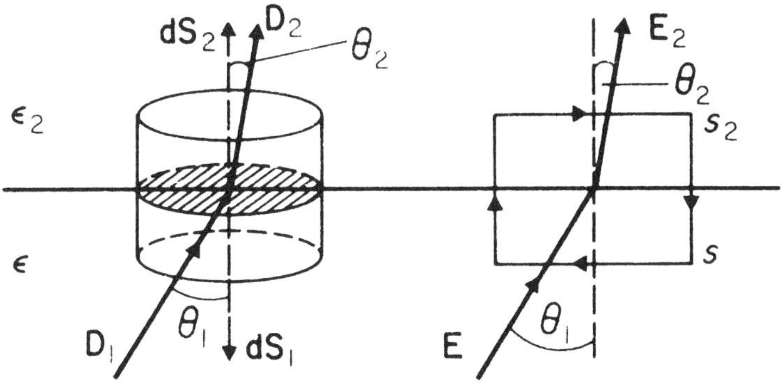

Fig. 2.22 Boundary conditions for the electric vectors $\boldsymbol{D}$ and $\boldsymbol{E}$ crossing between two dielectrics of permittivities ε_1 and ε_2.

At the boundary we therefore have continuity for D_n and E_t. When it is valid to use (2.32) we can write (2.33) and (2.34) as

$$\varepsilon_{r1} E_1 \cos\theta_1 = \varepsilon_{r2} E_2 \cos\theta_2$$

$$E_1 \sin\theta_1 = E_2 \sin\theta_2$$

We therefore get refraction of **D** and **E** at the boundary with the relation

$$\frac{\tan\theta_1}{\tan\theta_2} = \frac{\varepsilon_{r1}}{\varepsilon_{r2}} \tag{2.35}$$

2.5.4 Energy density

For a vacuum parallel plate capacitor we showed that the energy stored in it (2.20) is given by

$$U = \tfrac{1}{2} C V^2$$

With a dielectric completely filling it, the capacitance is increased by a factor ε_r (2.24) and the electric field is reduced by ε_r (2.28). Hence the energy stored is

$$U = \frac{1}{2}\left(\frac{\varepsilon_r \varepsilon_0 A}{d}\right)(Ed)^2$$

where E is the reduced field. Therefore the energy density is

$$u = \tfrac{1}{2}\varepsilon_r \varepsilon_0 E^2 = \tfrac{1}{2} D E$$

where (2.32) applies. As with the energy density in the vacuum (Fig. 2.17), the energy density still resides in the electric field. The difference is that (2.23) generally now becomes

$$U = \frac{1}{2}\int_\tau \boldsymbol{D}\cdot\boldsymbol{E}\,\mathrm{d}\tau \tag{2.36}$$

We can apply the principle of virtual work to the force between two capacitance plates (charged conductors) in a dielectric liquid (Fig. 2.18, with a dielectric present) and find

$$F = \frac{-\partial U}{\partial x} = \frac{-Q^2}{2}\frac{\partial}{\partial x}\left(\frac{1}{C}\right)$$

The dielectric increases C by a factor ε_r and so decreases F by $1/\varepsilon_r$. However this only leads to a revised Coulomb's law (compare (2.3)) in certain cases:

$$\boldsymbol{F}_1 = \frac{1}{4\pi\varepsilon_r\varepsilon_0} \sum_j \frac{q_1 q_j}{r_{1j}^2} \hat{\mathbf{r}}_{1j}$$

It is limited to dielectrics which are isotropic, homogeneous, linear and have a constant relative permittivity ε_r. In practice this limits it to fluids over a narrow range of temperature and pressure, whereas the vacuum version of Coulomb's law is always true for stationary charges.

3

Electric potential and fields

The electrostatic field $\boldsymbol{E}$ is a vector field which, by (2.17), is just the gradient of a scalar field ϕ, the electric potential. That is,

$$\boldsymbol{E} = -\operatorname{grad} \phi$$

The two most important properties of a vector field are its flux and its circulation. For the electrostatic field these are expressed by integral equations (2.9) and (2.10) as follows:

$$\text{Gauss's law} \qquad \int_S \boldsymbol{E}\cdot\mathbf{d}\boldsymbol{S} = \frac{1}{\varepsilon_0}\int_V \rho \, \mathrm{d}\tau$$

$$\text{Circulation law} \oint_C \boldsymbol{E}\cdot\mathbf{d}\boldsymbol{s} = 0$$

In this chapter theorems from vector analysis are used to derive the differential forms of these equations and hence to describe the electric potentials and fields for a number of cases of practical importance.

3.1 POISSON'S AND LAPLACE'S EQUATIONS

For an arbitrary vector field $\boldsymbol{F}(\boldsymbol{r})$ the flux out of a closed surface S enclosing a small volume $\mathrm{d}\tau$ is $\int_S \boldsymbol{F}\cdot\mathbf{d}\boldsymbol{S}$ where $\mathbf{d}\boldsymbol{S}$ is the outward normal from $\mathrm{d}S$ (compare Fig. 2.4). The flux of $\boldsymbol{F}$ per unit volume at the point $\boldsymbol{r}$ is called the *divergence* of $\boldsymbol{F}$ and is given by the limit of $\mathrm{d}\tau$ tending to zero:

$$\operatorname{div} \boldsymbol{F} = \operatorname*{Lim}_{\mathrm{d}\tau \to 0} \frac{\int_S \boldsymbol{F}\cdot\mathbf{d}\boldsymbol{S}}{\mathrm{d}\tau} \tag{3.1}$$

Gauss's divergence theorem follows from this definition and equates the total flux of a vector field $\boldsymbol{F}$ out of a closed surface S to the volume integral of the divergence of $\boldsymbol{F}$ over the volume V enclosed by S:

$$\int_S \boldsymbol{F}\cdot \mathbf{d}\boldsymbol{S} = \int_V \operatorname{div} \boldsymbol{F}\, \mathrm{d}\tau \tag{3.2}$$

The theorem applies to any vector field $\boldsymbol{F}(\boldsymbol{r})$ that is a smoothly varying field, that is $\boldsymbol{F}(\boldsymbol{r})$ is a continuously differentiable function of the coordinates of $\boldsymbol{r}$.

Applying this theorem to Gauss's law for the electrostatic field $\boldsymbol{E}$, we have

$$\int_S \boldsymbol{E}\cdot \mathbf{d}\boldsymbol{S} = \int_V \operatorname{div} \boldsymbol{E}\, \mathrm{d}\tau = \frac{1}{\varepsilon_0}\int_V \rho\, \mathrm{d}\tau$$

and so

$$\operatorname{div} \boldsymbol{E} = \rho/\varepsilon_0 \tag{3.3}$$

where ρ is the total electric charge density and ε_0 is the electric constant. Alternatively, from (2.31) we can write

$$\operatorname{div} \boldsymbol{D} = \rho_{\mathrm{f}} \tag{3.4}$$

where ρ_{f} is the volume density of the free charges. The divergence of $\boldsymbol{E}$ or $\boldsymbol{D}$ will be zero when their outward and inward fluxes for a volume V are equal and opposite. This corresponds to the net charge density in the volume V being zero. Equations (3.3) and (3.4) show that the divergence of a vector field is a scalar field.

The circulation of a vector field $\boldsymbol{F}(\boldsymbol{r})$ around a closed loop C is $\oint_C \boldsymbol{F}\cdot \mathbf{d}\boldsymbol{S}$ and the circulation per unit area of the loop $\mathrm{d}S$ is used to define the *curl* of $\boldsymbol{F}$. In general the circulation per unit area will depend on the orientation of $\mathrm{d}\boldsymbol{S}$ relative to $\boldsymbol{F}$ and so the curl of $\boldsymbol{F}$ is a vector quantity, which is defined in terms of its component (curl $\boldsymbol{F}$)$\cdot\hat{\mathbf{n}}$ normal to $\mathrm{d}S$ (Fig. 3.1), the direction of curl $\boldsymbol{F}$ being defined by

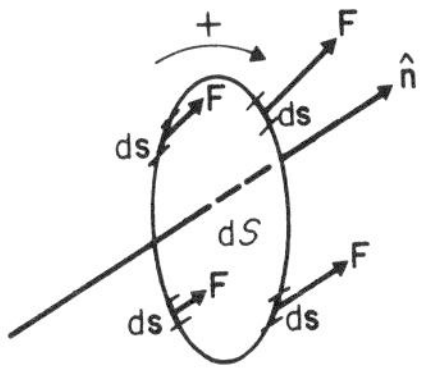

Fig. 3.1 Right-handed screw to evaluate curl $\boldsymbol{F}$.

a right-handed screw. Like the divergence, the curl of $\boldsymbol{F}$ is given by the limit of dS tending to zero:

$$(\text{curl } \boldsymbol{F})\cdot\hat{\mathbf{n}} = \lim_{\mathrm{d}S\to 0} \frac{\oint_C \boldsymbol{F}\cdot\mathbf{d}\boldsymbol{S}}{\mathrm{d}S} \tag{3.5}$$

Stokes's theorem follows from this definition and equates the circulation of a vector field $\boldsymbol{F}$ around a closed loop C to the flux of curl $\boldsymbol{F}$ through any surface S bounded by C:

$$\oint_C \boldsymbol{F}\cdot\mathbf{d}\boldsymbol{S} = \int_S (\text{curl } \boldsymbol{F})\cdot\mathbf{d}\boldsymbol{S} \tag{3.6}$$

Applying this theorem to the circulation law for the electrostatic field $\boldsymbol{E}$, we have the important result

$$\text{curl } \boldsymbol{E} = 0 \tag{3.7}$$

The fact that the electrostatic field is a curl-free field makes electrostatic problems the most straightforward ones to solve in electromagnetism.

In vector calculus the differential operator *del* is defined by

$$\nabla = \hat{\mathbf{i}}\frac{\partial}{\partial x} + \hat{\mathbf{j}}\frac{\partial}{\partial y} + \hat{\mathbf{k}}\frac{\partial}{\partial z} \tag{3.8}$$

in Cartesian coordinates, where $\hat{\mathbf{i}}, \hat{\mathbf{j}}, \hat{\mathbf{k}}$ are unit vectors along $0x, 0y, 0z$. In terms of this operator, the grad, div, and curl of an arbitrary scalar Ω and vector $\boldsymbol{F}(F_x, F_y, F_z)$ are as follows:

$$\text{grad } \Omega = \nabla\Omega = \hat{\mathbf{i}}\frac{\partial \Omega}{\partial x} + \hat{\mathbf{j}}\frac{\partial \Omega}{\partial y} + \hat{\mathbf{k}}\frac{\partial \Omega}{\partial z} \tag{3.9}$$

$$\text{div } \boldsymbol{F} = \nabla\cdot\boldsymbol{F} = \frac{\partial F_x}{\partial x} + \frac{\partial F_y}{\partial y} + \frac{\partial F_z}{\partial z} \tag{3.10}$$

$$\text{curl } \boldsymbol{F} = \nabla\times\boldsymbol{F} = \begin{vmatrix} \hat{\mathbf{i}} & \hat{\mathbf{j}} & \hat{\mathbf{k}} \\ \dfrac{\partial}{\partial x} & \dfrac{\partial}{\partial y} & \dfrac{\partial}{\partial z} \\ F_x & F_y & F_z \end{vmatrix} \tag{3.11}$$

The square of the del operator, $\nabla^2 = \nabla\cdot\nabla$, is a scalar operator of particular importance in electromagnetism, known as the *Laplacian*. Formally,

$$\nabla^2 = \left(\hat{\mathbf{i}}\frac{\partial}{\partial x} + \hat{\mathbf{j}}\frac{\partial}{\partial y} + \hat{\mathbf{k}}\frac{\partial}{\partial z}\right)\cdot\left(\hat{\mathbf{i}}\frac{\partial}{\partial x} + \hat{\mathbf{j}}\frac{\partial}{\partial y} + \hat{\mathbf{k}}\frac{\partial}{\partial z}\right)$$

$$= \frac{\partial^2}{\partial x^2} + \frac{\partial^2}{\partial y^2} + \frac{\partial^2}{\partial z^2} \tag{3.12}$$

and, in terms of the vector operators,

$$\nabla^2 \Omega = \mathrm{div}(\mathrm{grad}\ \Omega) \tag{3.13}$$

The scalar and vector fields resulting from the operations of ∇^2, grad, div and curl are here expressed in Cartesian coordinates, but their importance in vector analysis arises from their independence of any particular coordinate system. The latter can then be chosen according to the symmetry of the problem to which the differential equations are being applied.

For the electrostatic field, from (2.17) and (3.3) we have

$$\nabla^2 \phi = \mathrm{div}(\mathrm{grad}\phi) = \mathrm{div}(-\boldsymbol{E}) = -\rho/\varepsilon_0 \tag{3.14}$$

This is *Poisson's equation* for the electrostatic potential ϕ in the presence of an electric charge density ρ. It becomes *Laplace's equation*:

$$\nabla^2 \phi = 0 \tag{3.15}$$

in the absence of any charges.

3.2 SOLUTIONS OF LAPLACE'S EQUATION

The Laplacian operator in the spherical polar coordinates (r, θ, ψ) shown in Fig. 2.5 is

$$\frac{1}{r^2}\frac{\partial}{\partial r}\left(r^2 \frac{\partial}{\partial r}\right) + \frac{1}{r^2 \sin\theta}\frac{\partial}{\partial \theta}\left(\sin\theta \frac{\partial}{\partial \theta}\right) + \frac{1}{r^2 \sin\theta}\frac{\partial^2}{\partial \psi^2} \tag{3.16}$$

For the simplest solutions of Laplace's equation, let the electric potential ϕ be symmetrical about the polar axis so that $\partial\phi/\partial\psi = 0$. Then (3.15) and (3.16) give:

$$\frac{\partial}{\partial r}\left(r^2 \frac{\partial \phi}{\partial r}\right) + \frac{\partial}{\partial(\cos\theta)}\left\{(1-\cos^2\theta)\frac{\partial\phi}{\partial(\cos\theta)}\right\} = 0$$

For the highest symmetry ϕ is independent of both θ and ψ or

$$\frac{\partial}{\partial r}\left(r^2 \frac{\partial \phi}{\partial r}\right) = 0$$

and it is obvious (by substitution) that $\phi_1 = r^{-1}$ is a solution, where for a sphere $r^2 = x^2 + y^2 + z^2$ in Cartesian coordinates. Since ϕ_1 is a solution of Laplace's equation, then so are partial derivatives of

ϕ_1 with respect to the space coordinates, such as $\partial\phi_1/\partial x$, $\partial\phi_1/\partial z$. If the polar axis is $0z$ then $z = r\cos\theta$ (Fig. 2.5) and

$$\frac{\partial\phi_1}{\partial z} = -\frac{1}{r^2}\left(\frac{z}{r}\right) = -\frac{z}{r^3} = -\frac{\cos\theta}{r^2}$$

Hence $\phi_2 = r^{-2}\cos\theta$ is a solution and further solutions can be found be successive differentiation:

$$\frac{\partial\phi_2}{\partial z} = \frac{1}{r^3} - \frac{3}{r^4}\left(\frac{z^2}{r}\right) = \frac{1}{r^3}(1 - 3\cos^2\theta)$$

Thus $\phi_3 = r^{-3}(1 - 3\cos^2\theta)$ is a solution. In general it can be shown that the solutions are $\phi_n = r^n P_n$ and $\phi_{n+1} = r^{-(n+1)}P_n$, $n = 0, 1, 2\ldots$ where P_n are *Legendre* functions. In electrostatics $\phi_1 = r^{-1}$ is just the spherically symmetric potential of a point charge (2.13), while $\phi_2 = r^{-2}\cos\theta$ is the potential at a distance $r \gg s$ for a dipole ((2.26) and exercise 1) and $\phi_3 = r^{-3}(1 - 3\cos^2\theta)$ is the long-range potential of a linear quadrupole (exercise 3).

These solutions can also be applied to such problems as a conducting sphere (exercise 5) or a dielectric sphere (exercise 6) in a uniform electric field. In each case the electric potentials can be used to obtain the electric fields from the polar components of grad ϕ:

$$(\text{grad}\,\phi)_r = \frac{\partial\phi}{\partial r} \text{ and } (\text{grad}\,\phi)_\theta = \frac{1}{r}\frac{\partial\phi}{\partial\theta} \tag{3.17}$$

In cylindrical polar coordinates (r, ϕ, z) shown in Fig. 3.2, the Laplacian operator becomes

$$\frac{1}{r}\frac{\partial}{\partial r}\left(r\frac{\partial}{\partial r}\right) + \frac{1}{r^2}\frac{\partial^2}{\partial\phi^2} + \frac{\partial^2}{\partial z^2} \tag{3.18}$$

but it is important to note that r is now normal to the z-axis and no longer a position vector from the origin. For a long coaxial cable the cylindrical symmetry produces a radial $\phi(r)$, except at the ends. Hence

$$\nabla^2\phi = \frac{1}{r}\frac{\partial}{\partial r}\left(r\frac{\partial}{\partial r}\right)\phi(r) = 0$$

Since r is finite, on integration we obtain first

$$r\frac{\partial\phi(r)}{\partial r} = A$$

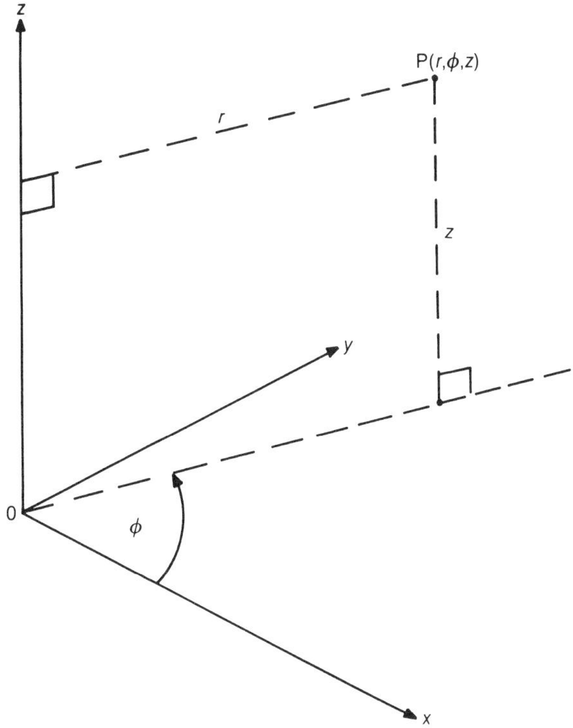

Fig. 3.2 Cylindrical polar coordinates (r, ϕ, z).

and second

$$\phi(r) = A \ln r + B \tag{3.19}$$

where A and B are constants.

For the cross-section of the cable shown in Fig. 3.3, the usual arrangement is for the outer conductor of inner radius b to be earthed and the inner conductor of outer radius a to be at a potential V. Hence (3.19) becomes

$$A \ln a + B = V$$

$$A \ln b + B = 0$$

Subtracting

$$A \ln b/a = -V$$

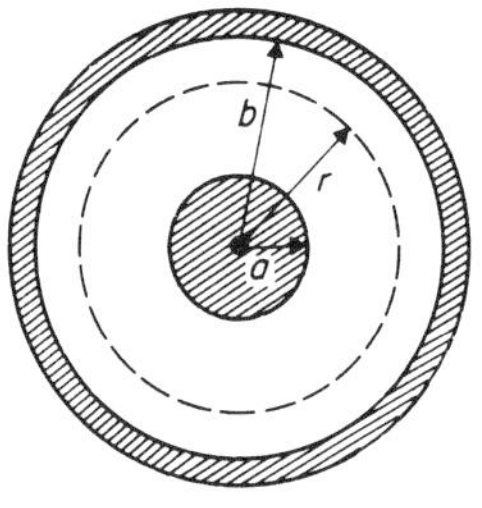

Fig. 3.3 Cross-section of a coaxial cable.

Hence (3.19) becomes

$$\phi(r) = \frac{-V}{\ln(b/a)}\{\ln r - \ln b\} \tag{3.20}$$

and the electric field

$$E_r = -\frac{\partial \phi}{\partial r} = \frac{-A}{r} = \frac{V}{r \ln(b/a)}$$

This agrees with the solution obtained from Gauss's law (Chapter 2 and exercise 13).

In equation (3.20) the potential $\phi(r)$ is a solution of Laplace's equation that satisfies the boundary conditions of the problem and is valid for $a \leqslant r \leqslant b$. Since we found this function in a systematic manner by integration of Laplace's equation, we know that it is the only function with these properties. In many practical problems, with less symmetry, it is not usually possible to integrate Laplace's equation directly. Sometimes a solution can be found by intuition, sometimes by using an analogue method. An important theorem, the *uniqueness theorem* (proved in formal treatises), states that any potential that satisfies both Laplace's (or Poisson's) equation and the appropriate boundary conditions for a particular electric field is the *only possible* potential. This applies to any arrangement of conductors and dielectrics and so is very useful.

3.3 ELECTRICAL IMAGES

The uniqueness theorem enables the electric potentials and fields of charge distributions to be found by a substitutional technique known as the method of images. In this method, for example, a conductor is replaced by a point 'image' charge such that the conducting surface is still an equipotential surface. By the uniqueness theorem the electric potentials for a point charge plus its image are identical with those for a point charge and the conductor for the region *outside* the surface, since in both cases Laplace's equation is satisfied for all points outside the conductor. The method is illustrated by some examples here and exercises 8–10.

3.3.1 Point charge and plane conductor

A simple example is a point charge $+q$ near an earthed, infinite conducting plane (Fig. 3.4(a)). The charge will induce a surface density of charge σ on the conductor, which is related to the electric field E_0 just outside the surface by

$$E_0 = \sigma/\varepsilon_0 \tag{3.21}$$

as we showed from Fig. 2.13. At the conductor the electric field is everywhere normal to it and the problem is to find the electric field distribution and hence the distribution of the reduced charge on the conductor. In this case a charge $-q$ placed at the position of a virtual image of $+q$ in the conducting plane (Fig. 3.4(b)) produces an equipotential surface (---) which is, like the conductor it replaces, at zero potential with respect to the two charges. Therefore the electric field produced at points such as P to the left of the zero potential surface by the charges $+q$ and $-q$ will be identical with

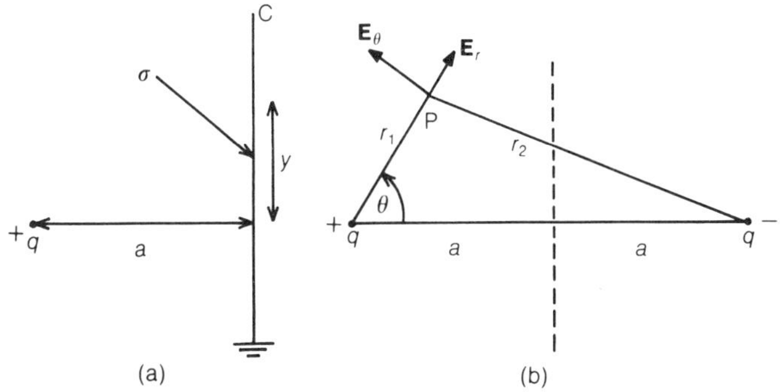

Fig. 3.4 (a) Point charge q at distance a from earthed conducting plane C induces surface charge density σ. (b) The image charge $-q$ at distance $2a$ from q replaces the conducting plane C.

that due to $+q$ and the earthed conductor, by the uniqueness theorem.

The potential ϕ_P at $P(r, \theta)$ is, from (2.13),

$$\phi_P = \frac{1}{4\pi\varepsilon_0}\left(\frac{q}{r_1} - \frac{q}{r_2}\right)$$

where, from Fig. 3.4(b),

$$r_2^2 = r_1^2 + 4a^2 - 4r_1 a \cos\theta$$

The polar components of the electric field are, from (2.17) and (3.17), given by

$$E_r = -\frac{\partial \phi_P}{\partial r} = \frac{1}{4\pi\varepsilon_0}\left\{\frac{q}{r_1^2} - \frac{q(r_1 - 2a\cos\theta)}{r_2^3}\right\} \tag{3.22}$$

and

$$E_\theta = -\frac{1}{r}\frac{\partial \phi_P}{\partial \theta} = -\frac{2qa\sin\theta}{4\pi\varepsilon r_2^3} \tag{3.23}$$

The electric field lines are plotted in Fig. 3.5, together with two sections of equipotential surfaces, which cut the field lines normally everywhere.

The electric field at the surface of the conductor is everywhere normal to it and is given by

$$E_0 = E_r\cos\theta - E_\theta\sin\theta \tag{3.24}$$

At the surface $r_1 = r_2$ and E_0 points into the conductor, as the

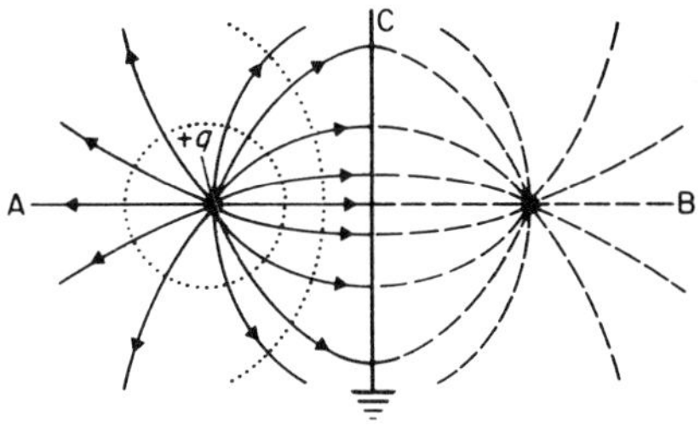

Fig. 3.5 Electric field lines (⟶) from the charge q near the conducting plane C. Sections through two equipotential surfaces are shown as dotted lines (······); the surfaces are generated by rotating those sections about AB. The imaginary field lines (----) of the image charge show the symmetry of the electric field lines, which are normal to the plane of C at all points on it.

induced charges are negative. From (3.21) to (3.24)

$$E_0 = \frac{2qa}{4\pi\varepsilon_0 r_1^3} = -\frac{\sigma}{\varepsilon_0}$$

and so

$$\sigma = \frac{-qa}{2\pi r_1^3}$$

As a check, if we calculate the total charge induced on the conductor, from Fig. 3.4(a) we obtain:

$$\int_0^\infty \sigma 2\pi y \mathrm{d}y = -\frac{q}{2}\int_0^\infty \frac{a\mathrm{d}(y^2)}{(a^2 + y^2)^{3/2}} = -q$$

which is just the image charge we placed at $-x$. All image charges for conductors are, as in this case, virtual images, on the opposite side of the conductor to the inducing charge.

Having found the correct image charge, it is also possible to use it for other field calculations. For example, the conductor in this example attracts the charge $+q$ with the same Coulomb force as that between $+q$ and its image $-q$, that is $-q^2/(4\pi\varepsilon_0 \cdot 4a^2)$. The solution to this example (Fig. 3.5) may also be useful in finding the answer to a similar problem. For example, the equipotential surfaces in Fig. 3.5 could be curved conductors and so this solution would also give the electric field outside a curved conductor, of the appropriate shape, when a positive charge is brought near to it. The fields associated with a spherical conductor are discussed in exercises 9 and 10.

3.3.1 Point charge and plane dielectric

The problem of the electric field surrounding a point charge q_1 near a plane dielectric (Fig. 3.6(a)) is more complex than that for a conductor since the field penetrates the dielectric. In this case two 'image' charges are suggested (Fig. 3.6(a)), q_2 at the virtual image position as before and q_3 outside the dielectric to represent the field inside. From Fig. 3.6(b), the potentials at P and Q are then

$$\phi_{\mathrm{P}} = \frac{1}{4\pi\varepsilon_0}\left(\frac{q_1}{r_1} + \frac{q_2}{r_2}\right)$$

$$\phi_{\mathrm{Q}} = \frac{q_3}{4\pi\varepsilon_0 r_3}$$

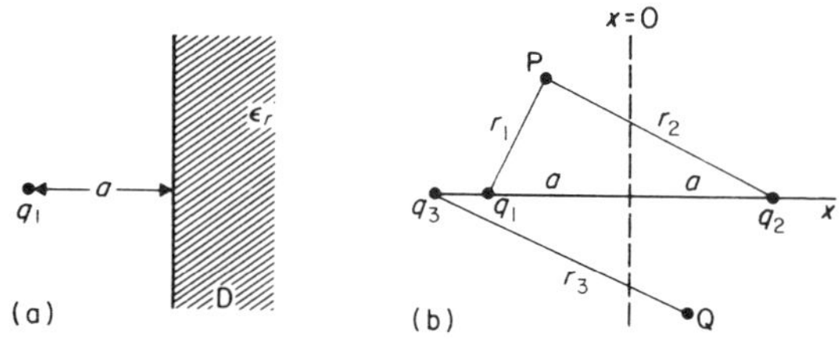

Fig. 3.6 (a) Point charge q at distance a from plane dielectric D of relative permittivity ε_r. (b) The image charge q_2 at distance $2a$ from q_1 for fields outside the dielectric and the image charge q_3 for fields inside the dielectric replace the dielectric material.

At the boundary $\phi_P = \phi_Q$ and, from (2.34), $E_{Pt} = E_{Qt}$. To satisfy these conditions we must have

$$r_1 = r_2 = mr_3$$

$$q_1 + q_2 = mq_3 \tag{3.25}$$

However, since $\phi_P = \phi_Q$ for *all* values of θ, $m = 1$. Therefore q_3 coincides with q_1 and is the image of q_2 in the surface.

The other boundary condition, from (2.33), is $D_{Pn} = D_{Qn}$ or

$$\frac{\partial \phi_P}{\partial x} = \varepsilon_r \frac{\partial \phi_Q}{\partial x}$$

since there is no free charge density at the surface of a dielectric. The analysis is essentially the same as in the previous example, the E_0 for a conductor becoming for the dielectric,

$$E_0 = \frac{2a(q_1 - q_2)}{4\pi\varepsilon_0 r_1^3} = \frac{\varepsilon_r \cdot 2aq_3}{4\pi\varepsilon_0 r_1^3}$$

and so

$$q_1 - q_2 = \varepsilon_r q_3 \tag{3.26}$$

The image charges, from (3.25) and (3.26), are therefore

$$q_2 = -\left(\frac{\varepsilon_r - 1}{\varepsilon_r + 1}\right) q_1$$

$$q_3 = \frac{2}{(\varepsilon_r + 1)} q_1$$

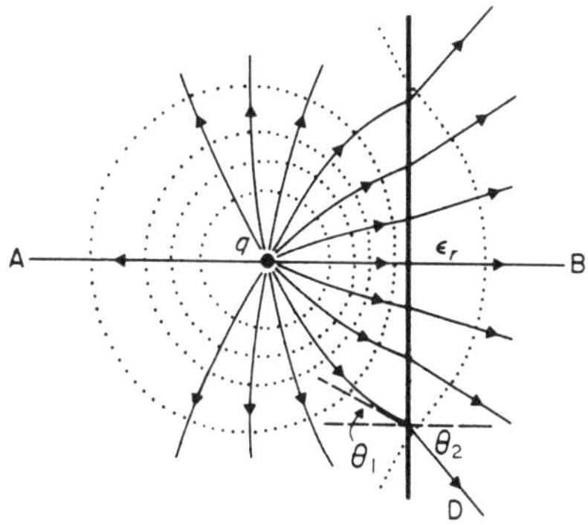

Fig. 3.7 Electric displacement lines (→) from the charge q near the plane dielectric D of relative permittivity ε_r. Sections through one equipotential surface in the dielectric and four outside it are shown (⋯⋯); their surfaces are generated by rotation about AB. Inside the dielectric the field lines are radial from q and the equipotential surfaces have spherical symmetry. At the boundary the field lines are refracted according to $\cot\theta_1 = \varepsilon_r \cot\theta_2$.

and the force of attraction of the chage q_1 to the dielectric is

$$F_1 = \frac{-q_1 q_2}{4\pi\varepsilon_0 \cdot 4a^2} = \frac{q_1^2(\varepsilon_r - 1)}{16\pi\varepsilon_0 a^2(\varepsilon_r + 1)}$$

In Fig. 3.7 the electric displacement $\boldsymbol{D}$ is described by the full lines and the equipotentials by the dotted lines. Inside the dielectric the electric field is due to the point charge q_3 and so is radial with spherical equipotential surfaces, but outside the induced surface charge superimposed on the point charge q_1 produces a complex field pattern. At the boundary, from (2.35),

$$\cot\theta_1 = \varepsilon_r \cot\theta_2$$

3.4 ELECTRON OPTICS

In electronic instruments with visual displays, such as oscilloscopes or electron microscopes, it is necessary to focus beams of electrons in high vacua. In such cases the electrons are accelerated from an electron gun by voltages of the order of kilovolts and emerge at non-relativistic speeds up to about 0.2c. On the other hand for particle physics experiments, or to produce synchrotron radiation, electrons can be accelerated to gigavolts to (10^9 eV) and the relativistic electron mass can be greater than that of a proton at rest. Electron optics is the study of electron trajectories over a wide range of speeds in vacuo and is a complex subject of great technical importance. Here only the principles of electrostatic lenses for non-relativistic electrons will be outlined, while magnetic lenses are described in the next chapter.

By analogy with geometrical optics, suppose we have a narrow beam of electrons travelling at velocity v_1 in an equipotential region at a potential of V_1 volts towards another equipotential region at a potential of V_2 volts (Fig. 3.8). At the boundary plane there is a potential gradient, or electric field, that will accelerate the electrons normal to the boundary, but will not change their velocity parallel to the boundary. The emergent rays will be 'refracted' from an incident angle θ_1 to an angle θ_2 and their velocity v_2 will be given by

$$v_1 \sin\theta_1 = v_2 \sin\theta_2$$

But $\frac{1}{2}mv_1^2 = eV_1$ and $\frac{1}{2}mv_2^2 = eV_2$ so that

$$\frac{\sin\theta_1}{\sin\theta_2} = \frac{v_2}{v_1} = \sqrt{\frac{V_2}{V_1}}$$

This is analogous to Snell's law, $n_1 \sin\theta_1 = n_2 \sin\theta_2$, and we see that the electron velocity corresponds to the refractive index of the medium.

In a similar way since a convex lens focuses light, so a curved equipotential surface should focus electrons. Uniform electric fields (for example, Fig. 2.19) have plane equipotential surfaces associated with them, so we require non-uniform electric fields to focus electrons. A simple way to produce focusing is to have a single pair (Fig. 3.9(a)) or a double pair (Fig. 3.9(b)) of apertures, each in a plane conductor. The field lines change smoothly from the high-field to the low-field region producing the required curved equipotentials.

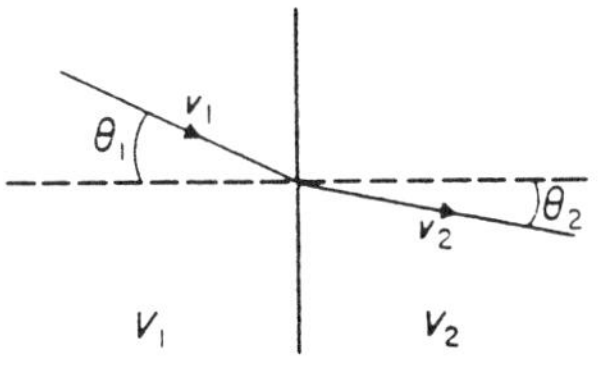

Fig. 3.8 Electrons at velocity v_1 in the equipotential region V_1 are refracted from an incident angle θ_1 to an emergent angle θ_2 at velocity v_2 in the equipotential region V_2.

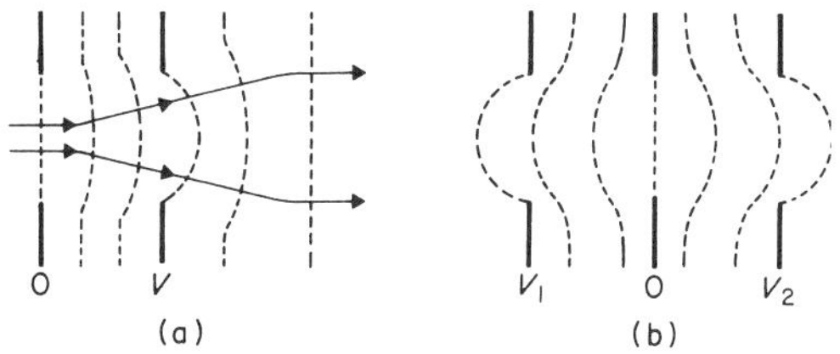

Fig. 3.9 Electric field lines (→) and equipotentials (----) for (a) a single pair, and (b) a double pair of apertures.

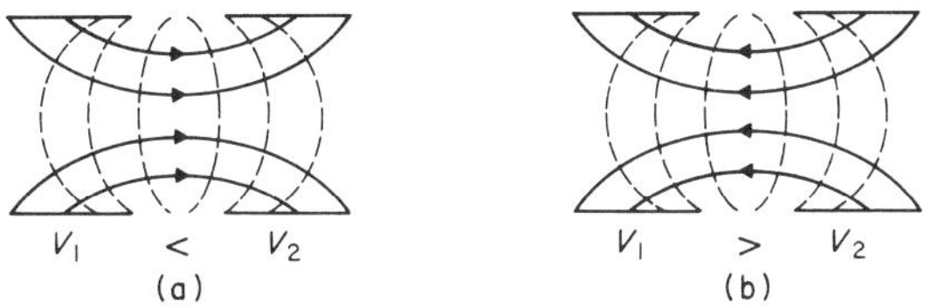

Fig. 3.10 Electric field lines (→) and equipotentials (----) for (a) an accelerating, and (b) a decelerating cylindrical lens.

Another method is to have a gap between two cylindrical electrodes (Fig. 3.10). It is interesting to note that both an accelerating lens (Fig. 3.10(a)) and a decelerating lens (Fig. 3.10(b)) are not converging lenses, since in each case the electrons spend more time in the converging regions. However, these 'simple' electrostatic lenses are not simple optically, for they both accelerate the electrons horizontally and act on them through the radial component of the electric field.

3.4.1 Thin electrostatic lens

In practice a lens of the type shown in Fig. 3.9(b) would have thick electrodes with rounded corners to avoid edge effects, as shown in Fig. 3.11(a). A system of this sort has cylindrical symmetry and so Laplace's equation in cylindrical polars, from (3.18), becomes

$$\frac{1}{r}\frac{\partial}{\partial r}\left(r\frac{\partial \phi}{\partial r}\right)+\frac{\partial^2 \phi}{\partial z^2}=0$$

Integrating

$$R\frac{\partial \phi}{\partial r}=-\int_0^R r\frac{\partial^2 \phi}{\partial z^2}\mathrm{d}r=-\frac{R^2}{2}\frac{\partial^2 \phi}{\partial z^2}$$

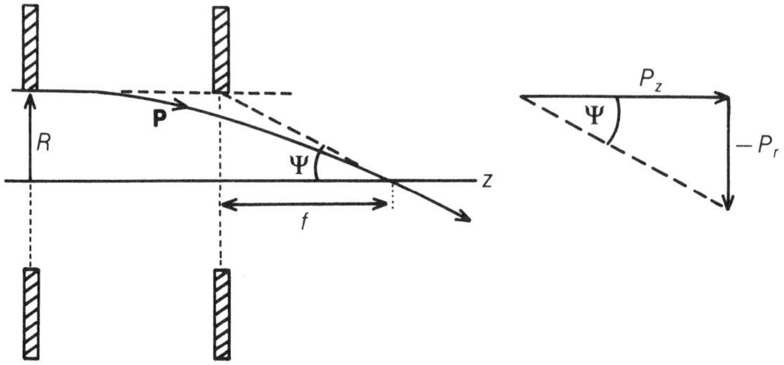

Fig. 3.11 (a) Paraxial electron path through an aperture showing the focal length of the lens. (b) Axial and radial electron moments, $\boldsymbol{p}_z$ and $\boldsymbol{p}_r$.

Thus

$$\frac{\partial\phi}{\partial r} = -\frac{R}{2}\frac{\partial^2\phi}{\partial z^2}$$

For electrons following paraxial trajectories, the axial velocity is $v_z = \mathrm{d}z/\mathrm{d}t$ at any point, while the radial momentum due to the radial force $e\partial\phi/\partial r$ on the electron is

$$p_r = \int e\frac{\partial\phi}{\partial r}\,\mathrm{d}t = -\frac{\mathrm{e}R}{2}\int\frac{\partial^2\phi}{\partial z^2}\frac{\mathrm{d}z}{v_z}$$

But the angle of deflection of ψ of each electron is, from Fig. 3.11, given by

$$\tan\psi = -\frac{p_r}{p_z} = -\frac{v_r}{v_z}$$

and, for paraxial rays, this is a small angle. We can therefore approximate v_z by the total speed $(2e\phi/m)^{1/2}$ at any point, so that

$$p_r = -\frac{\mathrm{e}R}{2}\cdot\left(\frac{m}{2e}\right)^{1/2}\int\frac{\partial^2\phi}{\partial z^2}\cdot\frac{\mathrm{d}z}{\phi^{1/2}}$$

This integral would be found numerically with a computer for a particular aperture and potential distribution, but it is evident that p_r is proportional to $-R$ for paraxial rays and so all the electrons with incident velocity parallel to the axis that pass through the aperture are brought to the same point on the axis. So there is a focus given by

$$\tan\psi = -\frac{p_r}{p_z} = \frac{R}{f}$$

Therefore the emergent focus for incident parallel rays from the central electrode in Fig. 3.9(b) is

$$\frac{1}{f_2} = \frac{\frac{1}{2}e(m/2e)^{1/2}}{(2eV_2m)^{1/2}}\int\frac{\partial^2\phi}{\partial z^2}\frac{\partial z}{\phi^{1/2}} = \frac{1}{4\sqrt{V_2}}\int\frac{\partial^2\phi}{\partial z^2}\frac{\partial z}{\phi^{1/2}}$$

A similar focus at $-f_1$ for electrons from the central electrode passing through V_1 (Fig. 3.9(b)), by symmetry, gives

$$\frac{f_1}{f_2} = -\sqrt{\frac{V_1}{V_2}}$$

which is an exact analogy of the thin lens formula $n_1/f_1 = -n_2/f_2$ in geometrical optics with refractive indices n_1 and n_2 on each side.

3.4.2 Electric quadrupole lens

As in optics electron lenses may cause aberrations, so that finite apertures can produce spherical aberrations and, if the incident electrons are not monoenergetic, they can produce chromatic aberrations. Unlike photons, electrons repel one another and so an incident parallel beam slowly diverges producing an electronic aberration.

An example of an astigmatic electron lens is a combination of two quadrupole lenses with alternate potentials, the first producing a

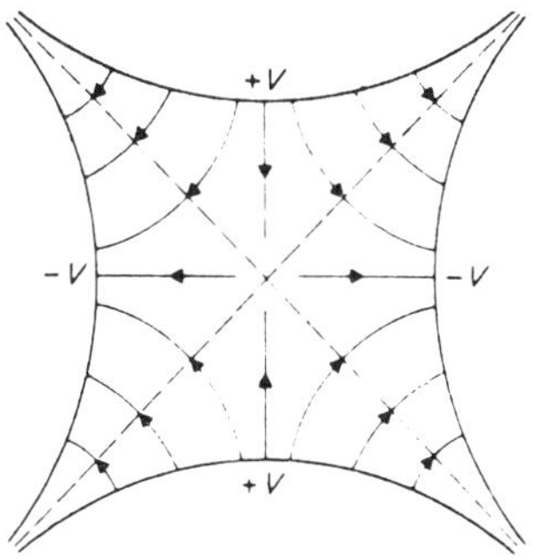

Fig. 3.12 Field lines (→) and hyperbolic conductors ($+V$, $-V$) for a cross-section of an electric quadrupole lens.

diverging focus and the second a converging focus. If the lenses have focal lengths $f_1 = -f_2 = f$ and are separated by a distance d, the combination has a focal length f^2/d, by analogy with a pair of thin optical lenses. Quadrupole lenses have axial, but not cylindrical, symmetry, as shown in Fig. 3.12. Laplace's equation then reduces to

$$\frac{\partial^2 \phi}{\partial x^2} + \frac{\partial^2 \phi}{\partial y^2} = 0$$

and ϕ is a two-dimensional potential, independent of the z-axis. In this example the equipotentials are $(x^2 - y^2)$ and the electric field lines from the hyperbolic conductors are $2xy$. As with the simple lens the electric field for an off-axis electron is proportional to the axial displacement (for example $-\partial\phi/\partial x = -2x$) and so produces a focus.

4

Magnetostatics

Electrostatics has been the study of electric fields, electric potentials and electric forces due to charges at rest, i.e. stationary charges. These charges were of two types: free charges on conductors and fixed charges, arising from polarization, on dielectrics. We concluded our study by considering how elecrostatic fields could be used to focus a diverging beam of electrons in a vacuum. In this chapter we first discuss the nature of a steady electric current I in matter and introduce the concept of the magnetic field $\boldsymbol{B}$, sometimes called the magnetic induction field. The relationships between steady or magnetostatic fields having $\partial \boldsymbol{B}/\partial t = 0$ and steady or direct electric currents having $\partial I/\partial t = 0$ are then developed and illustrated with examples.

4.1 ELECTRIC CURRENT

What is an electric current? The simplest idea is to think of a stream of electrons in a vacuum (Fig. 4.1(a)). If there are N electrons per unit volume each of charge $-e$, then the charge density in the

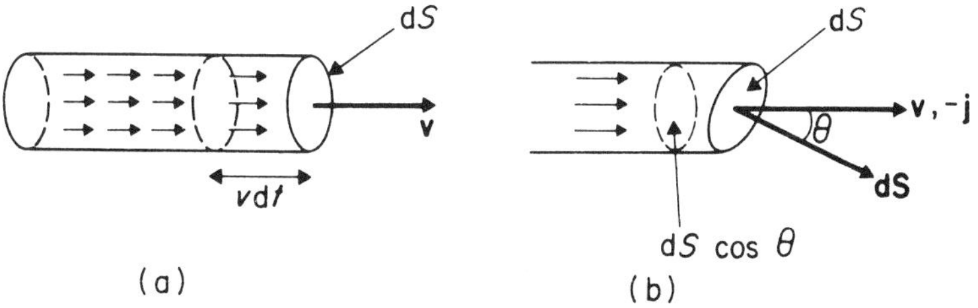

Fig. 4.1 (a) Electron stream in a vacuum. (b) Electric current is the flux of current density $\boldsymbol{j}\cdot\mathbf{dS}$.

vacuum, $\rho = -Ne$. If they are all moving with the same velocity $\boldsymbol{v}$, then the charge flowing out of the area dS in time dt, where dS is normal to $\boldsymbol{v}$, is

$$dq = -Ne\,dS \cdot v\,dt$$

Therefore, the charge flowing out per second is

$$\frac{dq}{dt} = -Nev \cdot dS$$

We define a *current density* vector $\boldsymbol{j}$ as the positive charge flowing per unit area per second at a point and so

$$\frac{dq}{dt} = j\,dS$$

when $\boldsymbol{j}$, and hence $\boldsymbol{v}$, is normal to dS. Otherwise (Fig. 4.1(b)) we write:

$$\frac{dq}{dt} = -Ne\boldsymbol{v}.\mathbf{dS} = \boldsymbol{j} \cdot \mathbf{dS}$$

where the scalar products $v dS \cos\theta$ and $j dS \cos\theta$ are similar to those found for the electric flux (Fig. 2.4). Hence

$$\boldsymbol{j} = -Ne\boldsymbol{v} = \rho\boldsymbol{v} \tag{4.1}$$

for a stream of electrons, where $q = -e$. An electric current I is measured for a particular area S and is therefore

$$I = \int_S \boldsymbol{j} \cdot \mathbf{dS} = \int \frac{dq}{dt} \tag{4.2}$$

Electric current is thus the flux of current density and has an SI unit of 1 ampere equal to 1 coulomb per second, while current density is in amperes per square metre.

Of course, electric currents commonly travel in copper wires, so let us try and estimate an electric current from our definition for uniform electron flow, $I = jS = -NevS$. Copper has one free electron per atom and about 6×10^{23} atoms per mole with an atomic mass of 63 and density 9 Mg m^{-3}. Therefore, $N = 6 \times 10^{23} \times 9 \times 10^6/63 = 10^{29}$ m^{-3}, $e = -1.6 \times 10^{-19}$ C and for a 1 mm diameter wire, $S \simeq 10^{-6}$ m^2. But what is a reasonable value for v? We might think that since a short pulse travels down a 10 m cable in about 1 μs that $v = 10^7$ m s^{-1}. Substituting this value, we get

$$I = 10^{29} \times 1.6 \times 10^{-19} \times 10^7 \times 10^{-6} \simeq 10^{11} \text{ A!}$$

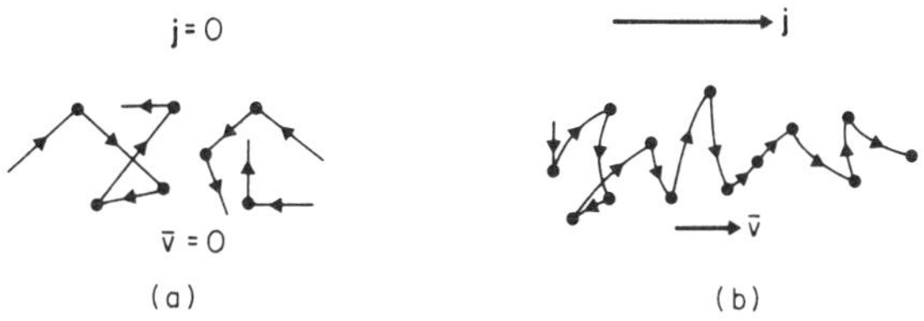

Fig. 4.2 Electron motion in a metal wire: (a) random, where there is no current flowing; (b) drifting in an electric current.

This is many orders of magnitude too big for a typical current, so v must be much smaller. In fact, for a current of 1 A, v will be only $10^{-4}\,\mathrm{m\,s^{-1}}$. This is extremely slow: it would take an electron in such a current about *three hours* to travel 1 metre.

This simple estimate shows the drastic difference between electrons in a vacuum and 'free' electrons in a copper wire. A copper wire with no current flowing in it has a total charge that is zero; overall it is electrically neutral. An idea of its microscopic state is shown in Fig. 4.2(a): there are exactly the same number of positive copper ions as there are 'free' negative electrons moving at random among them. When a constant electric field $\boldsymbol{E}$ is applied (Fig. 4.2(b)) each electron is accelerated during its free path by a force $-e\boldsymbol{E}$, but at each collision it loses its extra energy. The motion of the electrons through the wire is thus a diffusion process and we can associate a mean drift velocity $\bar{v}$ with this motion, which is just the acceleration times, the relaxation time, or, of the electrons' mean time τ between collisions. Thus,

$$\bar{v} = \left(\frac{-eE}{m}\right)\tau \tag{4.3}$$

where m is the electron mass. It is this drift velocity which we must use in (4.1) and not the much greater ($\sim 10^5\,\mathrm{m\,s^{-1}}$) speed of the electrons between their random collisions, so that

$$\boldsymbol{j} = Ne^2\boldsymbol{E}\tau/m \tag{4.4}$$

(A fuller discussion of this classical Drude theory of electrons in metals can be found in Professor Chambers' 'Electrons in metals and Semiconductors' in this series.)

It is found experimentally for most materials that the relaxation time τ of the electrons is independent of the applied field $\boldsymbol{E}$, but depends on such factors as the purity of the material and its temperature. We therefore define the *electrical conductivity* γ from the proportionality of $\boldsymbol{j}$ and $\boldsymbol{E}$.

$$\boldsymbol{j} = \gamma \boldsymbol{E} \tag{4.5}$$

which is Ohm's law, with

$$\gamma = Ne^2\tau/m \tag{4.6}$$

from (4.4). The elementary form of the law follows for a uniform wire of cross-section S and length l having a voltage V applied to it. Since $I = jS$ from (4.2) and $E = V/l$,

$$I = \gamma \frac{V}{l} S = \frac{V}{R} \tag{4.7}$$

where $R = l/\gamma S$ is the *resistance* of the wire of conductivity γ. The SI units are ohm (Ω) for R and (ohm metre)$^{-1}$ or siemens per metre (S m^{-1}) for γ. Typical values of conductivity are given in Table 4.1, showing the wide choice available from high-conductivity metals through semiconductors to insulators.

From (4.6) we see that the behaviour of γ (T), where T is the absolute temperature, depends on the number density N and τ. In metals N is independent of T and τ decreases as T rises, so that γ decreases with T rising. On the other hand, in semiconductors N increases as T rises, so that γ increases with rising T.

In a metal like copper it is easy to calculate the relaxation time τ from (4.6) using $\gamma = 6.0 \times 10^7$ S m^{-1} and the answer, $\tau = 2 \times 10^{-14}$ s,

Table 4.1 *Conductivity of materials at about 293 K*

Material	*Conductivity* (S m^{-1})
Copper	6.0×10^7
Silver	6.3×10^7
Lead	4.8×10^6
Mercury	1.0×10^6
Germanium	2.2
Silicon	1.8
Plate glass	3×10^{-11}
Mica	2×10^{-15}

Table 4.2 *Kinetic parameters for molecular and electron gases*

Parameter	*Air molecules at STP*	*Copper electrons at 293 K*	*Units*
Number density	4×10^{25}	10^{29}	m^{-3}
Typical drift velocity	1	10^{-4}	$m\,s^{-1}$
Relaxation time	2×10^{-10}	2×10^{-14}	s
Particle speed	500	10^5	$m\,s^{-1}$
Mean free path	100	2	nm

is an extremely short time. Since the conduction electrons in metals obey Fermi statistics, their random speed is their Fermi velocity, $v_F = 10^5\,m\,s^{-1}$. They therefore travel about $\tau v_F = 2$ nm, or five lattice spacings, on average between collisions. In Table 4.2 the relative kinetic parameters for molecules in air at STP and for conduction electrons in copper at 293 K are compared. We see that the electron gas is numerically much denser than air and the particles travel much faster, but they are scattered more frequently and so drift very slowly in a typical electric current.

4.2 LORENTZ FORCE

Having established what we mean by an electric current, we proceed to investigate the effects of currents on one another and on electric charges. Experiments such as those of Ampère in 1820 showed (Fig. 4.3) that parallel currents in adjacent wires attract one another, while antiparallel currents repel. As shown in Chapter 1, the wires themselves are electrically neutral to a very high degree (about 1 in 10^{25}), so the force acting is not an electrostatic or Coulomb force. Further experiments have shown (Fig. 4.4) that there is no force acting between a stationary charge and a wire carrying a current,

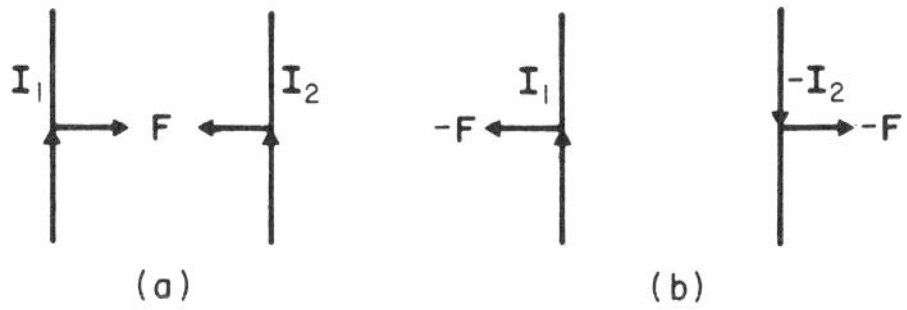

Fig. 4.3 Forces between currents: (a) parallel currents attract; (b) antiparallel currents repel.

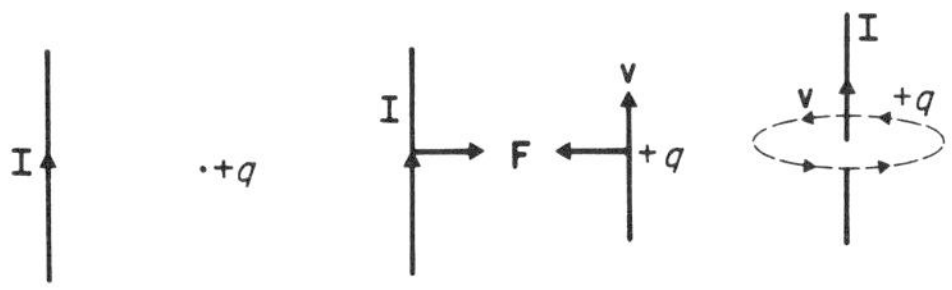

Fig. 4.4 Forces between a current and a charge. (a) No force for a stationary charge. (b) Attractive force for charge travelling parallel to the current. (c) No force for charge circulating in a plane normal to the current.

but charges moving parallel to a current do attract it, while charges circulating round in a plane normal to the current produce no effect.

Exactly similar effects are found to be produced when currents are placed between the poles of a permanent magnet (Fig. 4.5): the currents are repelled out of the magnet in a direction depending on the direction of the current. Today we describe all these effects by attributing a *magnetic field* $\boldsymbol{B}$ to *both* the current in the wire *and* the (atomic) currents in the magnet. Experiments on charges q moving at a velocity $\boldsymbol{v}$ in the uniform field $\boldsymbol{B}$ of a magnet then show that the forces $\boldsymbol{F}$ shown in Fig. 4.3 to 4.5 can all be explained in terms of a new force acting on a moving charge in a magnetic field called the *Lorentz force*. This force is proportional to the following:

1. the charge q,
2. the speed v, and
3. the sine of the angle between $\boldsymbol{v}$ and $\boldsymbol{B}$.

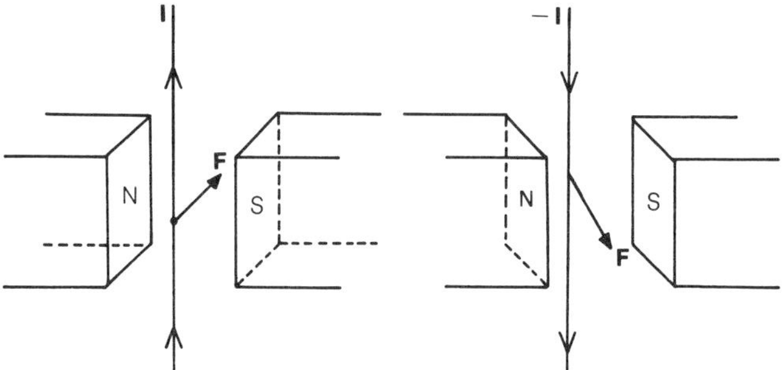

Fig. 4.5 Forces between a permanent magnet and a current. (a) Current upwards, force inwards. (b) Current downwards, force outwards.

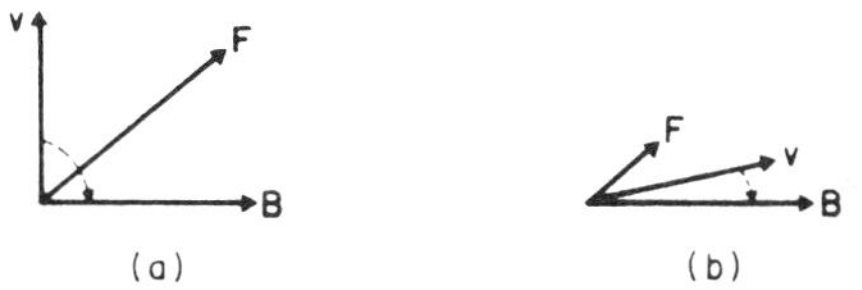

Fig. 4.6 The Lorentz force $\boldsymbol{F}$ on a charge with velocity $\boldsymbol{v}$ in a uniform magnetic field $\boldsymbol{B}$. (a) A large force. (b) A small force.

Furthermore it is always perpendicular to the plane containing $\boldsymbol{v}$ and $\boldsymbol{B}$. It is given by

$$\boldsymbol{F} \propto q\boldsymbol{v} \times \boldsymbol{B}$$

and illustrated in Fig. 4.6. In the SI system the constant of proportionality is unity and this equation can be used to define the unit of magnetic field $\boldsymbol{B}$: one tesla is that magnetic field which produces a force of 1 newton on a charge of 1 coulomb moving at 1 metre per second normal to the field. Therefore in SI units,

$$F = q\boldsymbol{v} \times \boldsymbol{B} \qquad (4.8)$$

Magnetic fields produced by powerful superconducting solenoids can be 10 tesla or more. These are huge when compared with the earth's magnetic field (about 100 microtesla), but dwarfed by the microscopic fields near nuclei which can be 10 kilotesla. An older unit of $\boldsymbol{B}$, the gauss, is still sometimes used; 1 tesla is exactly equal to 10 kilogauss.

In the presence of both an electric field and a magnetic field, the total force on a charge q is the sum of the Coulomb and Lorentz forces. That is

$$\boldsymbol{F} = q\boldsymbol{E} + q\boldsymbol{v} \times \boldsymbol{B} \qquad (4.9)$$

This is not immediately obvious, for Coulomb's force described the interaction of stationary charges. It is one of the remarkable properties of electric charge that it is invariant at all speeds, even relativistic ones, so that (4.9) is always true.

4.2.1 Applications to charged particles

The Lorentz force law is used to find the trajectories of charged particles moving in steady magnetic fields. We shall now look at some typical examples.

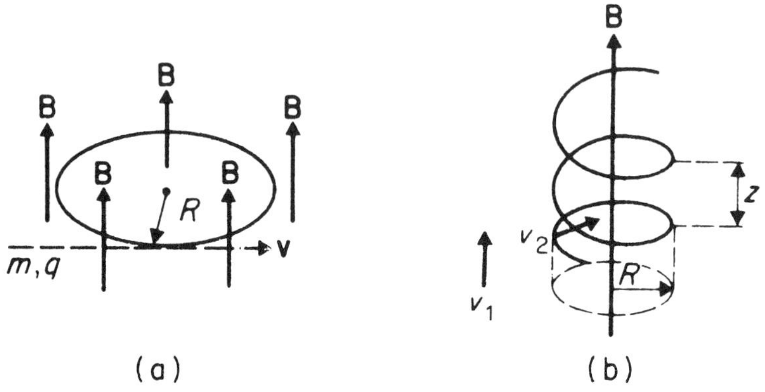

Fig. 4.7 Motion of a charged particle in a uniform magnetic field ***B***. (a) Enters field in plane normal to ***B***. (b) Enters field at another angle to ***B***.

1. *Circular motion*
When a charged particle enters a uniform magnetic field ***B*** with a velocity ***v*** in the plane normal to ***B***, the force $\boldsymbol{F} = q\boldsymbol{v} \times \boldsymbol{B}$ is constant and always normal to the motion. Therefore the trajectory is a circle of radius R (Fig. 4.7(a)) given by

$$F = \frac{mv^2}{R} = qvB$$

or

$$R = \frac{mv}{qB} \tag{4.10}$$

2. *Helical motion*
If the charged particle enters a uniform field ***B*** at any other angle, then we resolve ***v*** into v_1 parallel to ***B*** and v_2 normal to ***B*** (Fig. 4.7(b)). The speed parallel to ***B*** is not affected, while the speed normal to ***B*** would produce circular motion of radius $\boldsymbol{R} = mv_2/qB$. The total motion is therefore a helix around ***B*** of pitch $z = 2\pi m v_1/qB$.
3. *Magnetic quadrupole lens*
Non-uniform magnetic fields are used to focus beams of charged particles and so constitute magnetic lenses in television sets and particle accelerators. An arrangement of pole pieces similar to that of the electric quadrupole lens (Fig. 3.12) forms a magnetic quadrupole lens (Fig. 4.8(a)). It can be shown that an electron entering this magnetic field near the z-axis is deflected by the forces ***F*** to keep it in the horizontal xz-plane. The effect on a

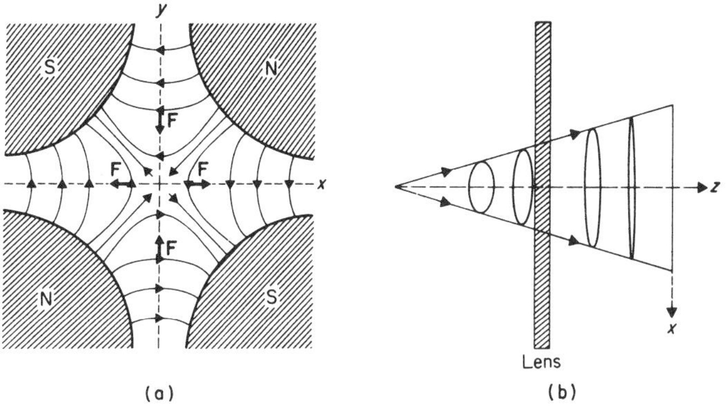

Fig. 4.8 (a) Magnetic field lines (→) and shaped pole pieces form a cross-section of a magnetic quadrupole lens. (b) The lens focuses a cone of electrons to a horizontal line.

cone of electrons is illustrated in Fig. 4.8(b) for this lens, while a similar lens with each polarity reversed would produce a line focus along $0y$. A pair of quadrupole lenses placed close together, or quadruple doublet, can therefore produce a point focus from a diverging beam.

4.2.2 Applications to electric currents

The Lorentz force law can also be used to investigate the motion of current carrying coils in magnetic fields.

1. *Current loop*
A coil carrying a current experiences a torque when placed in a uniform magnetic field due to the Lorentz forces acting on it. From Fig. 4.9(a) the current in side *bc* of the coil is, using (4.2),

$$I = -NevA$$

where A is the cross-sectional area of the wire and $\boldsymbol{v}$ is the drift velocity of the electrons of number density N. The Lorentz force acts on all these electrons and so is

$$\boldsymbol{F}_1 = -(NAL_1)e\boldsymbol{v} \times \boldsymbol{B} \tag{4.11}$$

with magnitude $F_1 = L_1IB$, and direction shown in Fig. 4.9(b). Similary for *ad*, $F_2 = L_1IB$, but acting in the opposite direction.

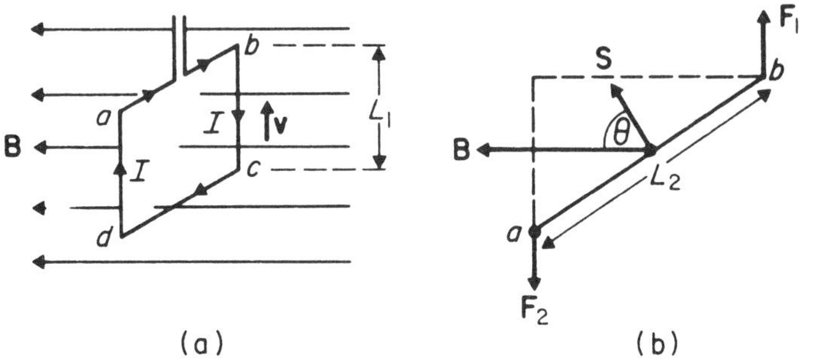

Fig. 4.9 Torque acting on a current loop in a magnetic field. (a) Side view. (b) Plan.

Another similar couple acts in the vertical plane on the sides *ab* and *dc*. The net force acting to displace the coil is therefore zero, but if the plane of the coil is not normal to $\boldsymbol{B}$ there will be a torque acting to rotate the coil into this position of magnitude

$$T = F_1 L_2 \sin\theta = I L_1 L_2 B \sin\theta$$

where θ is the angle, shown in Fig. 4.9(b), between the normal $\hat{\mathbf{n}}$ to the plane of the coil and $\boldsymbol{B}$. If $\boldsymbol{S}$ is the vector area $L_1 L_2 \hat{\mathbf{n}}$, then the torque is

$$\boldsymbol{T} = I\boldsymbol{S} \times \boldsymbol{B}$$

For a small coil of arbitrary shape and area dS, known as a *current loop*, this equation is still true. For such a current loop we define the *magnetic dipole moment* by

$$\boldsymbol{m} = I\,\mathbf{d}\boldsymbol{S} \qquad (4.12)$$

where the vector $\mathbf{d}\boldsymbol{S}$ points in the direction given by applying the right-hand screw rule to the direction of the current. The SI unit of magnetic moment is therefore A m^2 and the torque on a current loop in a uniform field is

$$\boldsymbol{T} = \boldsymbol{m} \times \boldsymbol{B} \qquad (4.13)$$

A similar torque $\boldsymbol{p} \times \boldsymbol{E}$ can be shown to act on an electric dipole in a uniform electric field (exercise 4, Chapter 3).

2. *Current element*

A short length dl of a wire carrying a current I is called a *current element*. The direction of the current is indicated by the vector length **d*l***, which is always in the opposite direction to the drift velocity of the electrons $\boldsymbol{v}$. The force on a current element is, from (4.11),

$$\mathbf{d}\boldsymbol{F} = -(NA\,\mathrm{d}l)e\boldsymbol{v} \times \boldsymbol{B}$$

or

$$\mathbf{d}\boldsymbol{F} = +(NAev)\,\mathbf{d}\boldsymbol{l} \times \boldsymbol{B}$$

The magnitude of the current $I = NAev$, therefore

$$\mathbf{d}\boldsymbol{F} = I\,\mathbf{d}\boldsymbol{l} \times \boldsymbol{B} \tag{4.14}$$

This is an important extension of the Lorentz force law to current elements, which will enable us to obtain the force between steady currents, when we know the magnetic field due to a current element.

4.3 BIOT–SAVART LAW

In Chapter 2 we showed how to calculate the electric field at a point 1 (Fig. 4.10(a)) due to a charge density distribution. We obtained equation (2.5),

$$\boldsymbol{E}(1) = \frac{1}{4\pi\varepsilon_0}\int_{\substack{\text{all}\\\text{space}}} \frac{\rho(2)\hat{\mathbf{r}}_{12}}{r_{12}^2}\,\mathrm{d}\tau_2$$

where $\rho(2)$ was the charge density at point 2 and $\int \mathrm{d}\tau$ was a triple integral over all space. This was based on Coulomb's law and the principle of superposition of electric fields.

In magnetostatics there is a similar integral which relates the magnetic field at a point 1 (Fig. 4.10(b)) due to the current in a circuit by integrating over the complete circuit. It is more complicated than the electrostatic case because it depends on moving charges and that

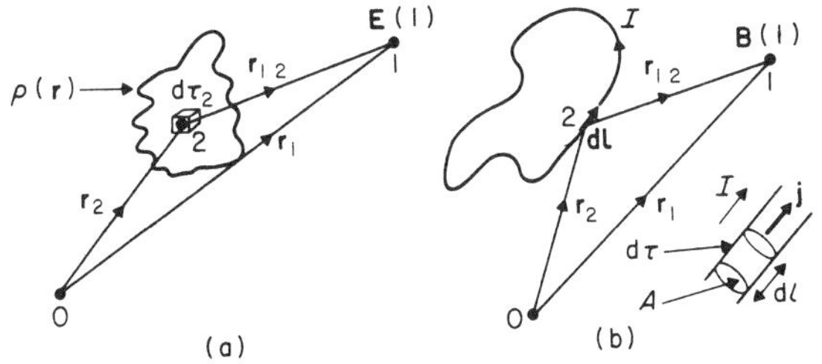

Fig. 4.10 (a) Electric field $\boldsymbol{E}(1)$ due to a distribution of charge. (b) Magnetic field $\boldsymbol{B}(1)$ due to a current.

implies a vector current density $\boldsymbol{j}(2)$ and a vector product between $\boldsymbol{j}(2)$ and $\hat{\mathbf{r}}_{12}$. The equation is

$$\boldsymbol{B}(1) = \frac{\mu_0}{4\pi} \int_{\substack{\text{all}\\ \text{space}}} \frac{\boldsymbol{j}(2) \times \hat{\mathbf{r}}_{12}}{r_{12}^2} \, \mathrm{d}\tau_2 \tag{4.15}$$

where the *magnetic constant*

$$\mu_0 = \frac{1}{\varepsilon_0 c^2} = 4\pi \times 10^{-7} \,\mathrm{N\,A^{-2}} \tag{4.16}$$

from equation (2.2).

If the circuit is composed of thin wires, then we can write

$$j \,\mathrm{d}\tau = jA \,\mathrm{d}l = I \,\mathbf{d}\boldsymbol{l}$$

where $\mathbf{d}\boldsymbol{l}$ is the vector length of a current element in the same direction as $\boldsymbol{j}$. Hence (4.15) becomes

$$\boldsymbol{B}(1) = \frac{\mu_0}{4\pi} \oint \frac{I \,\mathbf{d}\boldsymbol{l} \times \hat{\mathbf{r}}_{12}}{r_{12}^2} \tag{4.17}$$

where the integral is taken all round the circuit. This is the law named after Biot and Savart who, with Ampère, showed that it fitted the results of several experiments undertaken around 1820. Its status in magnetostatics is similar to that of Coulomb's law in electrostatics.

4.3.1 Applications of the Biot–Savart law

The Biot–Savart law is invaluable in calculating the magnetic fields due to simple circuits.

1. *Infinite wire*

   The magnitude of $\boldsymbol{B}$ due to an infinite wire carrying current I is, at point P (Fig. 4.11(a)):

$$B = \frac{\mu_0}{4\pi} \int_{-\infty}^{+\infty} \frac{I \,\mathrm{d}l \sin\theta}{r_{12}^2}$$

   To evaluate this integral, we note that the perpendicular distance R is fixed, that $r_{12} = R/\cos\phi$, $\sin\theta = \cos\phi$ and $l = R\tan\phi$. Hence the integral becomes

$$B = \frac{\mu_0 I}{4\pi} \int_{-\pi/2}^{\pi/2} R \sec^2\phi \,\mathrm{d}\phi \cos\phi \left(\frac{\cos^2\phi}{R^2} \right)$$

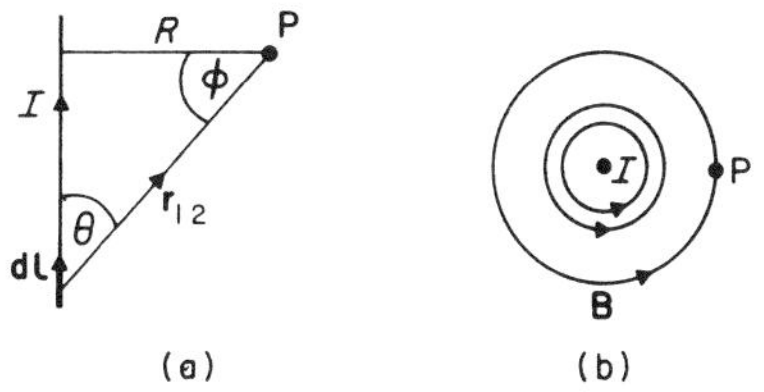

Fig. 4.11 Magnetic field at P due to an infinite wire. (a) Side view of current. (b) Plan of field lines.

and
$$B = \frac{\mu_0 I}{4\pi R}[\sin\phi]_{-\pi/2}^{\pi/2} = \frac{\mu_0 I}{2\pi R} \tag{4.18}$$

The direction of $\boldsymbol{B}$ is given by the right-hand screw rule for the vector product $\mathbf{d}\boldsymbol{l} \times \boldsymbol{r}_{12}$ and so is into the paper at P. Combining these results we see that $\boldsymbol{B}$ forms concentric circular field lines in planes normal to the wire (Fig. 4.11(b)) and falls off as $1/R$.

2. *Current loop*
A circular current around the origin in the xy-plane (Fig. 4.12) produces a field $\mathbf{d}\boldsymbol{B}$ normal to each current element $I\,\mathbf{d}\boldsymbol{l}$ and each distance vector $\boldsymbol{r}_{12}$. By symmetry the total $\boldsymbol{B}$ will be along the z-axis and be

$$B = \oint \mathrm{d}B_z = \frac{\mu_0}{4\pi}\oint \frac{I\,\mathrm{d}l\cos\theta}{r_{12}^2}$$

But $\cos\theta = a/r_{12} = a/(a^2 + z^2)^{1/2}$ is a constant and $\oint \mathrm{d}l = 2\pi a$, so that

$$B = \frac{\mu_0 I a^2}{2(a^2 + z^2)^{3/2}}$$

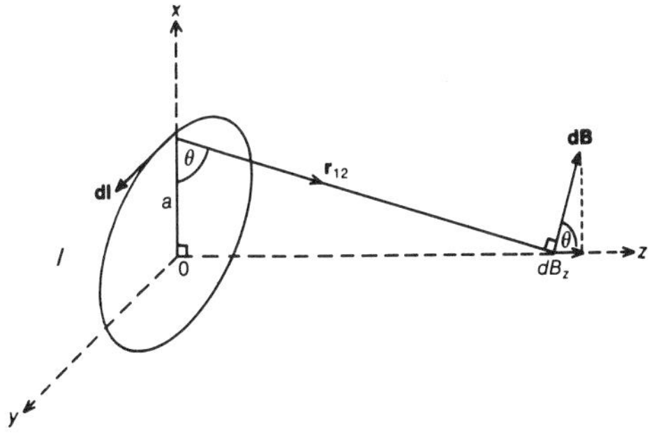

Fig. 4.12 Magnetic field on axis of a current loop.

The direction of $\boldsymbol{B}$ is towards you when you look into an anticlockwise (positive) current.

At the centre of the loop $\boldsymbol{B}$ has a maximum amplitude of $\mu_0 I/2a$, while at a large distance $z \gg a$,

$$B = \frac{\mu_0 I a^2}{2z^3} = \frac{\mu_0}{4\pi}\frac{2m}{z^3} \tag{4.19}$$

where $m = I\pi a^2$ is the magnetic dipole moment. Thus the field of a current loop at long range falls off as $1/z^3$ in a similar way to that of an electric dipole (exercise 1, Chapter 3).

4.4 FORCES BETWEEN CURRENTS

We can now combine (4.14), the Lorentz force law for current elements, with the Biot–Savart law (4.17) to obtain the forces between two current elements. In general, for Fig. 4.13(a), we have

$$\mathbf{d}\boldsymbol{B}_1 = \frac{\mu_0}{4\pi}\left(\frac{I_2\,\mathbf{d}\boldsymbol{l}_2 \times \hat{\mathbf{r}}_{12}}{r_{12}^2}\right)$$

and therefore from (4.14)

$$\mathbf{d}\boldsymbol{F}_1 = \frac{\mu_0}{4\pi}\left\{\frac{I_1\,\mathbf{d}\boldsymbol{l}_1 \times (I_2\,\mathbf{d}\boldsymbol{l}_2 \times \hat{\mathbf{r}}_{12})}{r_{12}^2}\right\} \tag{4.20}$$

and similarly

$$\mathbf{d}\boldsymbol{F}_2 = \frac{\mu_0}{4\pi}\left\{\frac{I_2\,\mathbf{d}\boldsymbol{l}_2 \times (I_1\,\mathbf{d}\boldsymbol{l}_1 \times \hat{\mathbf{r}}_{21})}{r_{12}^2}\right\} \tag{4.21}$$

The directions of the magnetic fields and forces are shown in the figure and clearly $\mathrm{d}F_1 = \mathrm{d}F_2$ in magnitude but not in direction.

We note that the forces are attractive, as was shown in experiments

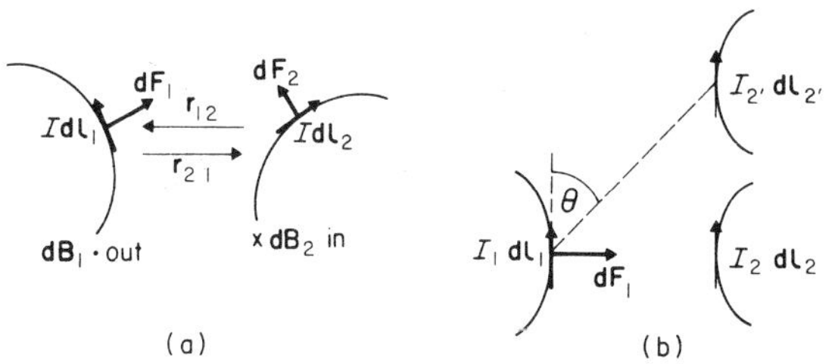

Fig. 4.13 Forces between currents: (a) random orientation; (b) parallel.

on parallel currents (Fig. 4.3(a)), and that there is apparently a net force when the wires are not parallel. However, there is no violation of Newton's third law, since on integrating these equations to get the total net force on a circuit the result is zero. For the simpler case of parallel wires (Fig. 4.13(b)), we obtain

$$\mathrm{d}F_1 = \frac{\mu_0 I_1\,\mathrm{d}l_1 I_2\,\mathrm{d}l_2 \sin\theta}{4\pi r_{12}^2}$$

when the current elements are displaced by an angle θ and

$$\mathrm{d}F_1 = \frac{\mu_0 I_1 I_2\,\mathrm{d}l_1\,\mathrm{d}l_2}{4\pi r_{12}^2} \tag{4.22}$$

when they are adjacent. The complexity of (4.20) and (4.21) arises directly from the Lorentz force law, and the apparent simplicity of (4.22) is due to the high symmetry of adjacent, parallel current elements.

The force between two currents is used in establishing the fourth base unit of the International System of Units (SI), that is the ampere: 'The ampere is that constant current which, if maintained in two straight parallel conductors of infinite length, of negligible circular cross-section, and placed 1 metre apart in vacuum, would produce between these conductors a force equal to 2×10^{-7} newton per metre of length'. This follows from (4.14), (4.16) and (4.18), which give the force per unit length required as

$$I \cdot \frac{\mu_0 I}{2\pi R} = 2 \times 10^{-7} \frac{I^2}{R}$$

4.5 AMPÈRE'S LAW AND MAGNETIC FLUX

In magnetostatics the circulation law corresponding to $\oint_C \boldsymbol{E}\cdot\mathbf{d}\boldsymbol{s} = 0$ in electrostatics is known as Ampère's law and enables us to find $\oint_C \boldsymbol{B}\cdot\mathbf{d}\boldsymbol{s}$. In general it is not zero and so $\boldsymbol{B}$ is *not* normally a conservative field, derivable from a scalar, magnetostatic potential. A simple example shows this to be true: the field lines around an infinite wire carrying a current I (Fig. 4.11) are circles in the planes normal to the wire. So if we evaluate $\oint_C \boldsymbol{B}\cdot\mathbf{d}\boldsymbol{s}$ around a circular path distance R from the wire, using (4.18) for $\boldsymbol{B}$, we have

$$\oint_C \boldsymbol{B}\cdot\mathbf{d}\boldsymbol{s} = \frac{\mu_0 I}{2\pi R}\oint_C \mathrm{d}s = \mu_0 I \tag{4.23}$$

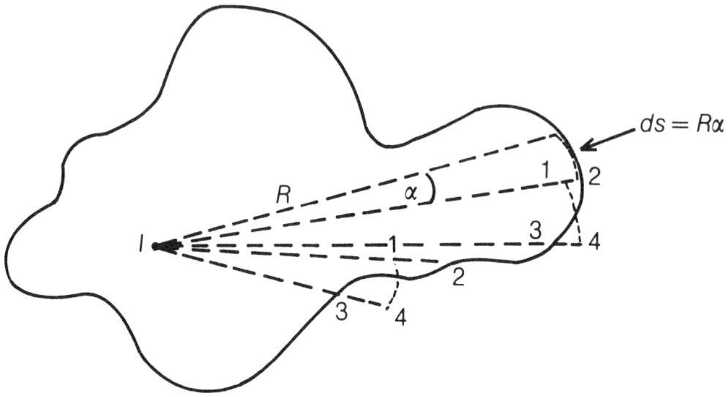

Fig. 4.14 Ampère's law applied to an irregular contour.

However, it is not necessary to choose a circular path for the contour, as Fig. 4.14 makes clear. Any irregular contour can be approximated by small arcs d$s = R\alpha$ plus radials (e.g. $1 \to 2$, $3 \to 4$), for each radial $\boldsymbol{B}$ due to I is normal to the radius vector and so

$$\oint_{\text{radials}} \boldsymbol{B} \cdot \mathbf{d}\boldsymbol{s} = 0$$

On the other hand, for each arc $\boldsymbol{B}$ is proportional to $1/R$ and so $\boldsymbol{B} \cdot \mathbf{d}\boldsymbol{s}$ depends only on the angles α. Therefore,

$$\oint_C \boldsymbol{B} \cdot \mathbf{d}\boldsymbol{s} = \oint_{\text{arcs}} \boldsymbol{B} \cdot \mathbf{d}\boldsymbol{s} = \frac{\mu_0 I}{2\pi} \oint \alpha = \mu_0 I$$

Of course if $\oint_C \boldsymbol{B} \cdot \mathbf{d}\boldsymbol{s}$ does not enclose any currents, then

$$\oint_C \boldsymbol{B} \cdot \mathbf{d}\boldsymbol{s} = 0$$

as in the electrostatic case. In general there will be contributions from each current enclosed, which in turn are given by (4.2), so that

$$\oint_C \boldsymbol{B} \cdot \mathbf{d}\boldsymbol{s} = \mu_0 \int_S \boldsymbol{j} \cdot \mathbf{d}\boldsymbol{S} \tag{4.24}$$

This is *Ampère's law*: the circulation of $\boldsymbol{B}$ is μ_0 times the current density flux enclosed.

4.5.1 Applications of Ampère's law

Ampère's law, (4.23) or (4.24), is useful in obtaining the magnetic field of large, symmetrical circuits.

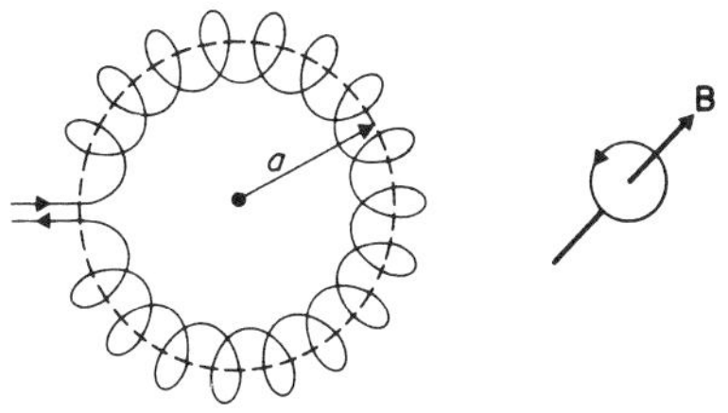

Fig. 4.15 (a) Ampèrian contour for a torus. (b) Magnetic field of each turn.

1. *Torus*
A torus (Fig. 4.15(a)) of N turns each carrying a current I has a uniform magnetic field B at the centre of each turn (Fig. 4.15(b)) and so a suitable Ampèrian contour is the circular axis of the torus. For this contour the currents on the inside thread the circle, while those on the outside do not. Therefore,

$$\oint_C \boldsymbol{B}\cdot\mathbf{d}\boldsymbol{s} = \mu_0 NI$$

from which

$$B = \frac{\mu_0 NI}{2\pi a} \tag{4.25}$$

2. *Long solenoid*
If we ignore the diverging fields at the ends of the long solenoid, then the uniform field $\boldsymbol{B}$ inside the solenoid can be found from Ampère's law. In Fig. 4.16 it is obvious that the large contour contains the current in each turn once and integrates the uniform field $\boldsymbol{B}$ over the length L of the solenoid. Therefore,

$$\oint_C \boldsymbol{B}\cdot\mathbf{d}\boldsymbol{s} = BL = \mu_0 NI \tag{4.26}$$

and

$$B = \mu_0 nI \tag{4.27}$$

where n is the number of turns per unit length. On the other hand a contour that does not enclose a current has zero circulation and so cannot be used to find $\boldsymbol{B}$ external to the solenoid. It is important to note that the field lines of $\boldsymbol{B}$ are always closed loops with no beginning or end point: this is because magnetic monopoles do not exist (compare Fig. 4.16 and 2.19). (Recent

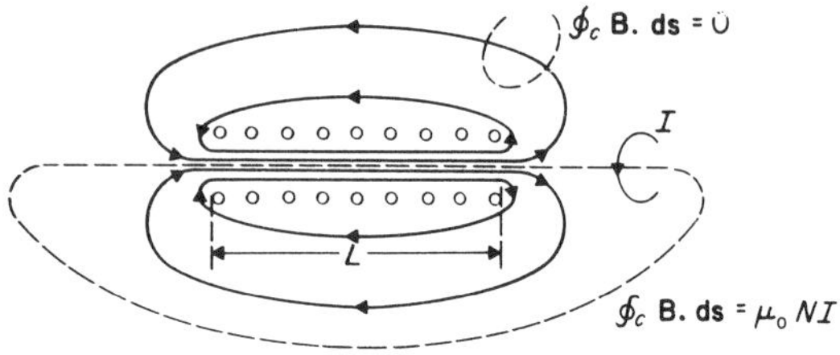

Fig. 4.16 Magnetic field lines and Ampèrian contours for a long solenoid.

theories of elementary particles imply that magnetic monopoles should exist in certain circumstances and so experimenters are trying to detect these rare particles.)

4.5.2 Magnetic flux

We define *magnetic flux*, Φ as $\int_S \boldsymbol{B}\cdot\mathbf{d}\boldsymbol{S}$, by analogy with the electric flux of Gauss's law. The SI unit of magnetic flux is the weber (Wb) and so the unit of magnetic field (or flux density) is sometimes taken as Wb m^{-2} rather than tesla. For any closed surface S

$$\Phi = \int_S \boldsymbol{B}\cdot\mathbf{d}\boldsymbol{S} = 0 \tag{4.28}$$

since there are no magnetic monopoles to make the total flux finite. This is *Gauss's law* for magnetism.

The concept of magnetic flux is mainly used for surfaces that are not closed, as illustrated in Fig. 4.17 for uniform and non-uniform fields. The *flux linkage* for a circuit of N turns is $N\,\Phi$ in each case.

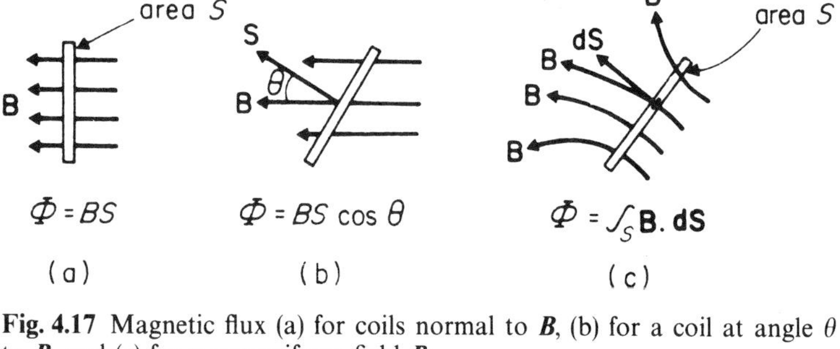

Fig. 4.17 Magnetic flux (a) for coils normal to $\boldsymbol{B}$, (b) for a coil at angle θ to $\boldsymbol{B}$, and (c) for a nonuniform field $\boldsymbol{B}$.

Applying Gauss's divergence theorem (3.2) to Gauss's law for magnetostatics (4.28), we see that

$$\int_S \boldsymbol{B}\cdot\mathbf{d}\boldsymbol{S} = \int_V \operatorname{div}\boldsymbol{B}\,\mathrm{d}\tau = 0$$

Therefore,

$$\operatorname{div}\boldsymbol{B} = 0 \tag{4.29}$$

and we see that $\boldsymbol{B}$ is always a divergence free field. Since this is due to the absence of magnetic monopoles, it is always true.

5

Electromagnetism

So far we have been able to consider electric fields and magnetic fields separately, through imposing the conditions that these fields and any currents shall be in steady states with $\partial \boldsymbol{E}/\partial t$, $\partial \boldsymbol{B}/\partial t$ and $\partial \boldsymbol{j}/\partial t$ all zero. In this chapter we move from steady currents to varying currents, from steady electric fields to induced electric fields, from stationary to moving circuits. Electric and magnetic effects become intimately connected in the study of electromagnetism, which has already been introduced in the electromagnetic force equation (4.9) as

$$\boldsymbol{F} = q\boldsymbol{E} + q\boldsymbol{v} \times \boldsymbol{B}$$

We first develop the concepts contained in Faraday's law and then apply it to a variety of examples.

5.1 FARADAY'S LAW

In a series of experiments in 1831–2 Faraday showed conclusively that electricity from batteries and magnetism from iron magnets were not separate phenomena but intimately related. He discovered that voltages can be generated in a circuit in three different ways:

1. by moving the circuit in a magnetic field;
2. by moving a magnet near the circuit; and
3. by varying the current in an adjacent stationary circuit.

Consider the simple system shown in Fig. 5.1: a solenoid producing an axial field $\boldsymbol{B}$ which passes through a current loop connected to a galvanometer (current detector). The galvanometer needle kicks if:

1. the solenoid is moved backwards and forwards;

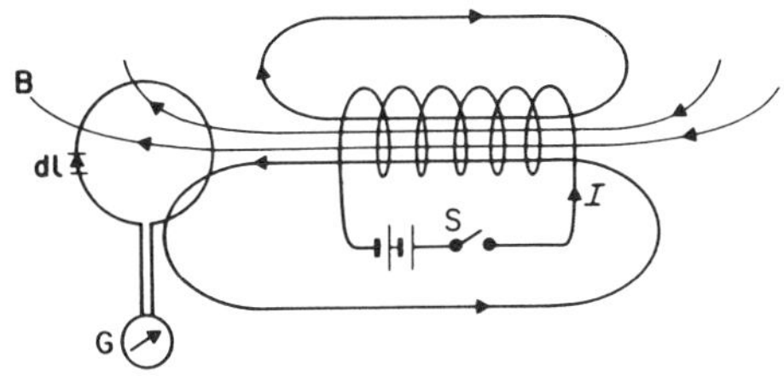

Fig. 5.1 A solenoid produces a magnetic field $\boldsymbol{B}$ which is sensed by a current loop and a galvanometer G.

2. the current loop is moved;
3. the current in the solenoid is switched on or off without moving any circuits.

The needle only moves when there is a current in the loop, i.e. when there is a net force on the electrons in the wire in one direction along it. There may be several different forces acting on different parts of the loop but what moves the needle is the net force integrated around the complete circuit. This is the *electromotive force* (e.m.f.)

$$\mathscr{E} = \oint \frac{\boldsymbol{F}}{q} \cdot \mathbf{d}\boldsymbol{l} = \oint \boldsymbol{E} \cdot \mathbf{d}\boldsymbol{l} \tag{5.1}$$

where $\boldsymbol{F}$ is the force on charge q and the integral is taken round the loop. The definition of $\mathscr{E}$ is therefore the tangential force per unit charge in the wire integrated over the complete circuit.

We can illustrate this definition by considering the work done by a battery of e.m.f. $\mathscr{E}$ driving a charge q around a circuit (Fig. 5.2). The work done by the battery is $\mathscr{E}q$, while that done on the charge in moving a distance $\mathbf{d}\boldsymbol{l}$ is $\boldsymbol{F} \cdot \mathbf{d}\boldsymbol{l}$. Therefore,

$$\mathscr{E}q = \int_a^b \boldsymbol{F} \cdot \mathbf{d}\boldsymbol{l}$$

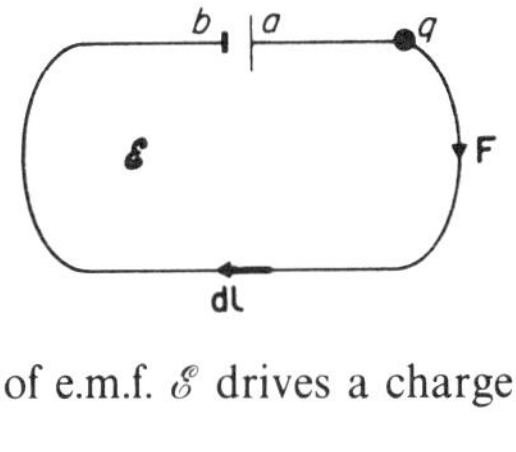

Fig. 5.2 A battery of e.m.f. $\mathscr{E}$ drives a charge q around a circuit.

and
$$\mathscr{E} = \int_a^b \frac{\boldsymbol{F}}{q} \cdot \mathbf{d}\boldsymbol{l} = \int_a^b \boldsymbol{E} \cdot \mathbf{d}\boldsymbol{l}$$

where $\boldsymbol{E} = \boldsymbol{F}/q$ is the electric field or force per unit charge. The SI units for $\mathscr{E}$ and $\boldsymbol{E}$ are obviously not the same: e.m.f. is measured in volts, electric field in volts per metre.

5.1.1 Motional e.m.f.s

The concept of electromotive force generated by the motion of circuits is best understood by considering some examples.

1. *Metal rod in a uniform field*
   In Fig. 5.3 a metal rod AB of length L is placed on the y-axis and moved along $0x$ in the uniform magnetic field $\boldsymbol{B}$ in the z-direction. By the Lorentz force law, each electron in the wire experiences a force $\boldsymbol{F} = -e\boldsymbol{v} \times \boldsymbol{B}$ and so the free electrons tend to move towards A. This produces a distribution of excess negative charge, which by Gauss's law is equivalent to an electric field $\boldsymbol{E}$ and so a net force $-e\boldsymbol{E}$ on each electron. Therefore,
   $$\boldsymbol{E} = \boldsymbol{v} \times \boldsymbol{B} \tag{5.2}$$
   From (5.1) the total e.m.f. across the rod will be
   $$\mathscr{E}_{\mathrm{AB}} = \int_{\mathrm{B}}^{\mathrm{A}} \boldsymbol{E} \cdot \mathbf{d}\boldsymbol{y} = v_x B_z L$$
   This e.m.f. is due to the electric field $\boldsymbol{E}$ induced in the rod by its motion through the operation of the Lorentz force law.
2. *Metal rod on rails in a uniform field*
   In Fig. 5.4 the same metal rod AB is mounted on metal rails so that a circuit ABCD of variable size is formed as the rod moves.

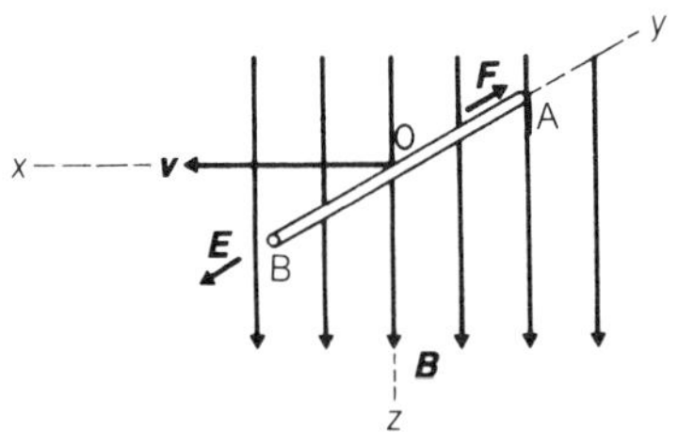

Fig. 5.3 Metal rod moving in a magnetic field.

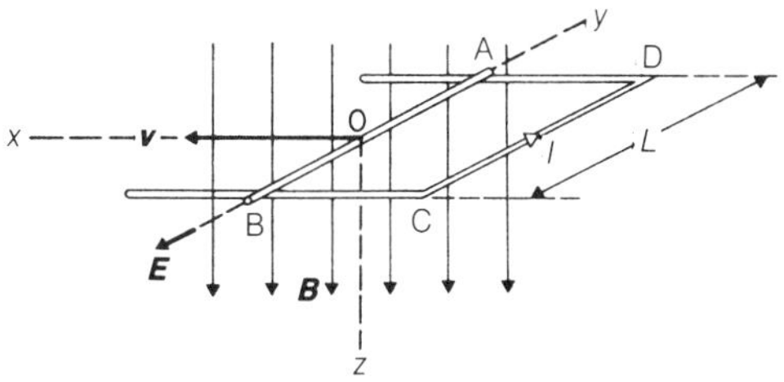

Fig. 5.4 A metal rod moving in a uniform magnetic field generates a current in a circuit.

The electrons now drift from B to A and go round DCB to form a conventional current I in the opposite direction. The e.m.f. generated by the rod moving is unchanged, but it now produces a current given by

$$I = \frac{v_x B_z L}{R}$$

where R is the total resistance of the circuit ABCD.

3. *Current loop in a uniform field*
   In Fig. 5.5 a square current loop abcd is moving into and out of a uniform magnetic field $\boldsymbol{B}$. If the sides of the loop are parallel to 0x and 0y and $\boldsymbol{B}$ is in the z-direction, then motion at velocity $\boldsymbol{v}$ along 0x will produce an e.m.f. $\mathscr{E}$ along ab as it enters the field. This will generate a current $I = v_x B_z l/r$, where l is the length ab and r is the resistance of the loop, until the side cd enters the field. Then an e.m.f. $\mathscr{E}$ will be generated in dc and this will exactly cancel that in ab giving zero current. Finally, as it emerges from the field there will only be the e.m.f. $\mathscr{E}$ in dc and this will generate I in the opposite direction in the loop.

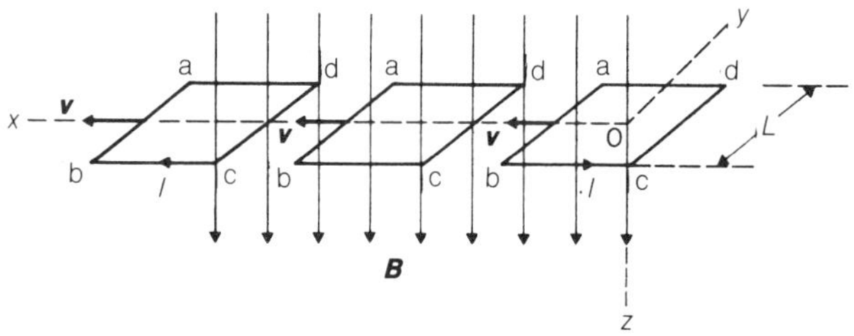

Fig. 5.5 Current loop moving in a uniform magnetic field.

The source of this electrical energy is the mechanical work done in moving the coil. It is dissipated in the loop as heat, which by Joule's law is I^2r watts. The mechanical work can equally well be done by moving the magnetic field across a stationary loop: it is essentially a relative motion effect. An observer on the coil in this case would see a moving magnetic field $\boldsymbol{B}$ for a time t as it passes him. As the coil enters and leaves $\boldsymbol{B}$, the observer sees a current in his stationary loop, which he ascribes to an electric field $\boldsymbol{E}$. This phenomenon of the transformation of electric and magnetic fields from one frame of reference to another in relative motion is described in Chapter 8. In terms of magnetic flux, at any instant the flux through the coil

$$\Phi = B_z l x$$

and so

$$|\mathscr{E}| = v_x B_z l = \frac{\mathrm{d}x}{\mathrm{d}t} B_z l = \frac{\mathrm{d}\Phi}{\mathrm{d}t} \tag{5.3}$$

This equation is an expression of the *flux rule* found experimentally by Faraday: an e.m.f. is induced in a circuit whenever the flux through it changes from any cause. The *direction* of the e.m.f. was established by Lenz and is known as *Lenz's law*: the current induced tends to oppose the change of flux through the circuit. The combination of the flux rule and Lenz's law is known as *Faraday's law*: the e.m.f. induced in a circuit is equal to the negative rate of change of the magnetic flux through that circuit. That is

$$\mathscr{E} = \oint \boldsymbol{E}\cdot\mathbf{d}\boldsymbol{l} = -\frac{\mathrm{d}\Phi}{\mathrm{d}t} \tag{5.4}$$

From the definition of magnetic flux this can be written

$$\oint \boldsymbol{E}\cdot\mathbf{d}\boldsymbol{l} = -\frac{\mathrm{d}}{\mathrm{d}t}\int_S \boldsymbol{B}\cdot\mathbf{d}\boldsymbol{S} \tag{5.5}$$

Here the direction of the vectors $\mathbf{d}\boldsymbol{S}$ are given by the right-hand screw rule for the circulation around S in the line integral.

Clearly the induced electric field in (5.5) is not an electrostatic field, for which the circulation law gives $\oint \boldsymbol{E}\cdot\mathbf{d}\boldsymbol{l} = 0$, but arises from the Lorentz force $q\boldsymbol{v} \times \boldsymbol{B}$ in the case of motional e.m.f.s and from $\mathrm{d}B/\mathrm{d}t$ when the magnetic field is varying. These can be different phenomena, although both are represented by Faraday's law. It is therefore important to distinguish between them.

5.1.2 Motional and transformer e.m.f.s

We can summarize the results on *motional e.m.f.s* by

$$\oint \boldsymbol{E}\cdot\mathbf{d}\boldsymbol{l} = \oint (\boldsymbol{v} \times \boldsymbol{B})\cdot\mathbf{d}\boldsymbol{l} \tag{5.6}$$

where $\boldsymbol{v}$ is the relative motion of a circuit with respect to the frame (usually the laboratory frame) in which $\boldsymbol{B}$ is fixed. We say that the circuit, which must not be changing in its shape or composition, is cutting the magnetic flux. This is true whether $\boldsymbol{B}$ is a steady magnetic field or a time-varying magnetic field.

The *transformer e.m.f.s* arise when there is no motion and $\boldsymbol{E}$ and $\boldsymbol{B}$ are fixed in the same coordinate system, so that

$$\oint \boldsymbol{E}\cdot\mathbf{d}\boldsymbol{l} = -\frac{\mathrm{d}}{\mathrm{d}t}\int_S \boldsymbol{B}\cdot\mathbf{d}\boldsymbol{S} = -\int_S \frac{\partial \boldsymbol{B}}{\partial t}\cdot\mathbf{d}\boldsymbol{S} \tag{5.7}$$

where the induced field $\boldsymbol{E}$ is due solely to the time variation of the magnetic field $\boldsymbol{B}$ and is therefore zero for a steady field. The time derivative now refers to each elementary area separately and so is a partial derivative. *Now there is no longer any flux cutting and so there is no need to restrict the line integral to a circuit.* It can be any contour in space and (5.7) becomes

$$\oint_C \boldsymbol{E}\cdot\mathbf{d}\boldsymbol{s} = -\int_S \frac{\partial \boldsymbol{B}}{\partial t}\cdot\mathbf{d}\boldsymbol{S} \tag{5.8}$$

Combining this with Stokes's theorem (3.6), we obtain the differential form of Faraday's law:

$$\operatorname{curl} \boldsymbol{E} = -\frac{\partial \boldsymbol{B}}{\partial t} \tag{5.9}$$

This is always true, giving as its time-independent limit the circulation law of the electrostatic field, $\operatorname{curl} \boldsymbol{E} = 0$. Equation (5.9) and the electromagnetic force equation (4.9) are the two fundamental equations of electromagnetism.

5.2 APPLICATIONS OF FARADAY'S LAW

We will now illustrate the usefulness of Faraday's law, as expressed in (5.4), (5.6) and (5.8).

5.2.1 Betatron

The betatron is a circular electron accelerator (Fig. 5.6) with the electrons circulating in a vacuum chamber placed in a powerful, non-uniform magnetic field produced by shaped pole-pieces. The electrons are accelerated by increasing the magnetic field, which generates an e.m.f. in the vacuum given by

$$\mathscr{E} = \oint_C \boldsymbol{E}\cdot\mathbf{d}\boldsymbol{s} = -\int_S \frac{\partial \boldsymbol{B}}{\partial t}\cdot\mathbf{d}\boldsymbol{S} \tag{5.10}$$

If we assume the electrons are injected into an orbit radius R for which the mean field is $\bar{B}$, then (5.10) becomes

$$2\pi R E = -\frac{\mathrm{d}\bar{B}}{\mathrm{d}t}\cdot\pi R^2$$

or

$$E = -\frac{R}{2}\frac{\mathrm{d}\bar{B}}{\mathrm{d}t}$$

The e.m.f. generated opposes the increasing magnetic flux and so its direction is clockwise when viewed from above the magnet (Fig. 5.6(b)) and therefore the electron motion is anticlockwise when driven by the force $-e\boldsymbol{E}$. The rate of change of momentum p of the

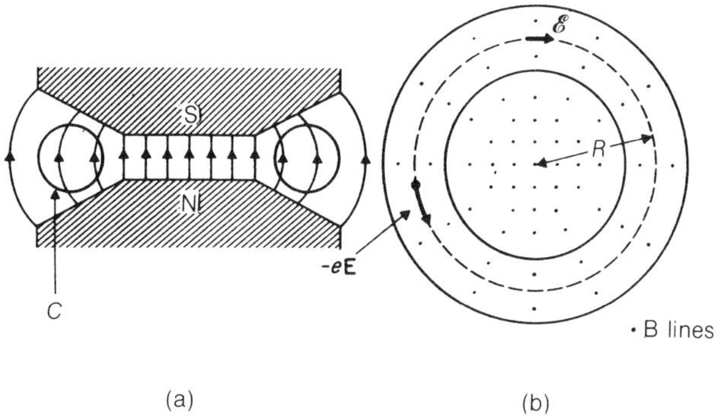

Fig. 5.6 The betatron. (a) Vertical section through the magnet NS and vacuum chamber C. (b) Top view of the vacuum chamber.

electron is therefore

$$\frac{\mathrm{d}p}{\mathrm{d}t} = -eE = \frac{eR}{2}\frac{\mathrm{d}\bar{B}}{\mathrm{d}t}$$

But we have already seen, in (4.10), that the momentum of an electron in a circular orbit is

$$p = eRB_R$$

where B_R in this case is the magnetic field at radius R. Clearly B_R must vary in time so that

$$\frac{\mathrm{d}p}{\mathrm{d}t} = \frac{eR\,\mathrm{d}B_R}{\mathrm{d}t} = \frac{eR}{2}\frac{\mathrm{d}\bar{B}}{\mathrm{d}t}$$

that is B_R must increase so that it is always equal to $\frac{1}{2}\bar{B}$ if the electrons are to be confined to their orbit as they accelerate. This is the principle of the betatron, which can accelerate electrons up to energies of many MeV, when they begin to radiate significantly.

5.2.2 Faraday's disc

A homopolar generator can be made from a disc rotating in a steady magnetic field, as first shown by Faraday. In Fig. 5.7 the circular disc of radius a rotates at a steady angular velocity ω in a uniform field $\boldsymbol{B}$. The simplest type of disc is an insulating one with a conducting ring round its circumference, a conducting axle RO and a radial conducting wire OP embedded in it. The circuit QROPQ is then completed by the brushes on the moving parts at Q and R. Since

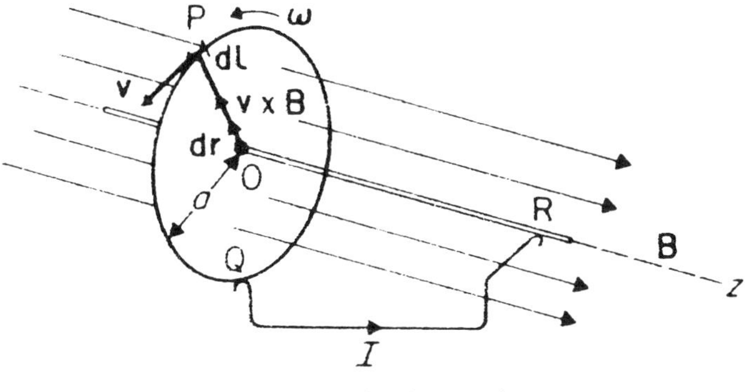

Fig. 5.7 Faraday's disc is a homopolar generator.

there is a steady field the only source of e.m.f. must be a motional e.m.f. given by (5.6). That is

$$\mathscr{E} = \oint \boldsymbol{E}\cdot\mathbf{d}\boldsymbol{l} = \oint (\boldsymbol{v} \times \boldsymbol{B})\cdot\mathbf{d}\boldsymbol{l}$$

The circuit QRO is stationary in the frame of $\boldsymbol{B}$ and so the only contributions to the e.m.f. come from OP and PQ. For OP, $\boldsymbol{v} \times \boldsymbol{B}$ is along $\boldsymbol{r}$ and $\mathbf{d}\boldsymbol{l} = \mathbf{d}\boldsymbol{r}$, while $v = r\omega$, so that

$$\oint (\boldsymbol{v} \times \boldsymbol{B})\cdot\mathbf{d}\boldsymbol{l} = \int_0^a r\omega B\,\mathrm{d}r = \frac{1}{2}a^2\omega B$$

And for PQ, $\boldsymbol{v} \times \boldsymbol{B}$ is normal to $\mathbf{d}\boldsymbol{l}$ at all points so that the contribution to the integral is zero. Therefore $\mathscr{E} = \frac{1}{2}a^2\omega B$ is the e.m.f. driving a current along the circuit in the direction QR.

In terms of Faraday's law (5.4), we must be careful to specify the flux being cut by the circuit. For the circuit in Fig. 5.7, the flux is cut as the radial wire OP sweeps round through angle POQ. If this angle is θ radians, then the flux cut $= B\pi a^2(\theta/2\pi)$ and Faraday's law gives

$$\mathscr{E} = -\frac{\mathrm{d}\Phi}{\mathrm{d}t} = -\frac{Ba^2}{2}\frac{\mathrm{d}\theta}{\mathrm{d}t} = +\frac{1}{2}a^2\omega B$$

as before, since $\omega\,\mathrm{d}t = -\mathrm{d}\theta$.

When the whole disc is a conductor (exercise 4), very high currents can be generated in the small resistance of the disc and so several brushes are joined together to conduct the radial current back to the axle.

5.2.3 Mutual inductance

A typical example of a transformer e.m.f. (5.8) is provided by a coil, 1, producing a time-varying magnetic field within the turns of a second coil, 2, fixed to it (Fig. 5.8(a)). When the same $\boldsymbol{B}$ is parallel to the area S of each of N turns of a coil, the e.m.f. becomes

$$\mathscr{E} = -\frac{\mathrm{d}}{\mathrm{d}t}(NBS)$$

that is, it is the negative of the rate of change of the flux linkage $N\Phi$.

Applying Ampère's law, $\oint \boldsymbol{B}\cdot\mathbf{d}\boldsymbol{s} = \mu_0 I$, to each turn of coil 1 in

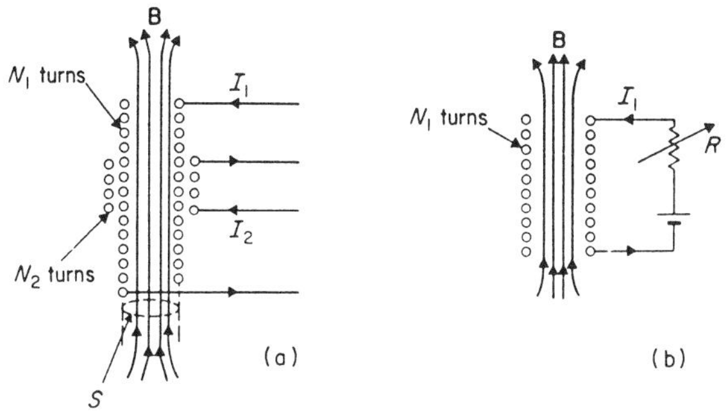

Fig. 5.8 (a) Mutual inductance. (b) Self-inductance.

Fig. 5.8(a), the field inside is

$$B_1 = \mu_0 N_1 I_1 / l_1$$

and for a long solenoid, B_0 outside is negligible. Therefore, the flux linking coil 2 from coil 1 is

$$N_2 \Phi_{21} = N_2 B_1 S = \mu_0 N_1 N_2 S I_1 / l_1$$

where I_1 is the only quantity varying with time. Hence the e.m.f. induced in coil 2 is

$$\mathscr{E}_{21} = -\left(\frac{\mu_0 N_1 N_2 S}{l_1}\right)\frac{\mathrm{d}I_1}{\mathrm{d}t}$$

Clearly $\mathscr{E}_{21}$ is proportional to $\mathrm{d}I_1/\mathrm{d}t$. The constant of proportionality, which depends only on the geometry of the coils, is called the *mutual inductance*, *M*. In particular M_{21} is the flux linking coil 2 due to *unit* current in coil 1. That is

$$M_{21} = \frac{N_2 \Phi_{21}}{I_1}$$

and

$$\mathscr{E}_{21} = -M_{21}\frac{\mathrm{d}I_1}{\mathrm{d}t} \tag{5.11}$$

For two overwound coils of the same length there is clearly a reciprocal relationship and the e.m.f. $\mathscr{E}_{12}$ induced in coil 1 due to current I_2 varying is

$$\mathscr{E}_{12} = -M_{12}\frac{\mathrm{d}I_2}{\mathrm{d}t} \tag{5.12}$$

where

$$M_{12} = \frac{N_1 \Phi_{12}}{I_2} \tag{5.13}$$

For this case it is obvious that $M_{12} = M_{21} = M$, a result which can be proved for any pair of coupled circuits (Neumann's theorem).

5.2.4 Self-inductance

Faraday induction is not confined to pairs of circuits: we can have self-induction in a single coil. In Fig. 5.8(b) if we vary R then I_1 changes and so the flux Φ_{11} changes. This generates a self-e.m.f. given by

$$\mathscr{E}_{11} = -\frac{\mathrm{d}}{\mathrm{d}t}(N_1 \Phi_{11})$$

The *self-inductance* of a coil L is therefore the flux linkage in the coil per unit current in the coil, or

$$L = \frac{N_1 \Phi_{11}}{I_1} \tag{5.14}$$

and

$$\mathscr{E} = -L\frac{\mathrm{d}I_1}{\mathrm{d}t} \tag{5.15}$$

The SI unit for both mutual and self-inductance is the *henry*, equivalent to 1 volt second per ampere. Since $M_{21} = \mu_0 N_1 N_2 S/l_1$ we note that an alternative, and commonly used, unit for the magnetic constant μ_0 is henry per metre, so that

$$\mu_0 = 4\pi \times 10^{-7}\,\mathrm{H\,m^{-1}} \tag{5.16}$$

an exact relationship. The self-inductance of a coil is

$$L = M_{11} = \mu_0 N^2 S/l \tag{5.17}$$

and an inductance of 1 H requires a very large, air-cored solenoid (e.g. 7000 turns of diameter 0.1 m over a length 0.5 m). Commonly air-cored coils are in the range μH to mH and larger ones have cores of high permeability to enhance their value. At very high frequencies (> 10 MHz) the 'skin effect' causes a small change in L, but for most purposes the self-inductance of an air-cored coil given by (5.17) is independent of frequency.

5.2.5 Coupled circuits

When two circuits are coupled, as in Fig. 5.9(a), the current in 1 depends on both the battery V_1 and the e.m.f. induced in it by variations in the current in 2. From (5.12),

$$\mathscr{E}_{12} = -M_{12}\frac{dI_2}{dt} = -M\frac{dI_2}{dt}$$

where the negative sign means that if both I_1 and I_2 are positive (anticlockwise) currents, then $\mathscr{E}_{12}$ will be opposed to V_1 when dI_2/dt is positive (increasing). If the polarity of V_1 is reversed, then I_1 is negative and so $\mathscr{E}_{12}$ and V_1 act in the same direction for increasing dI_2/dt.

The mutual inductance is reversed in sign if one of the coils is reversed, as can be seen more clearly in Fig. 5.9(b), where the total inductance will be

$$L_{\text{ABCD}} = L_1 + L_2 + 2M$$

or

$$L_{\text{ABDC}} = L_1 + L_2 - 2M \tag{5.18}$$

according to whether the currents in L_1 and L_2 are both of the same, or of opposite, signs. So, in general, the e.m.f. in circuit 1 of Fig. 5.9(a) is

$$V_1 \pm M\frac{dI_2}{dt} - L_1\frac{dI_1}{dt} = I_1 R$$

where the e.m.f. $\mathscr{E}_{11}$ from (5.15) has also been included.

When the coils L_1 and L_2 are coupled tightly together, for example by winding one on top of the other in a toroid, there is no leakage of magnetic flux and so

$$\Phi_{12} = \Phi_{22} \quad \text{and} \quad \Phi_{21} = \Phi_{11}$$

But

$$M_{12} = \frac{N_1\Phi_{12}}{I_2} \quad \text{and} \quad M_{21} = \frac{N_2\Phi_{21}}{I_1}$$

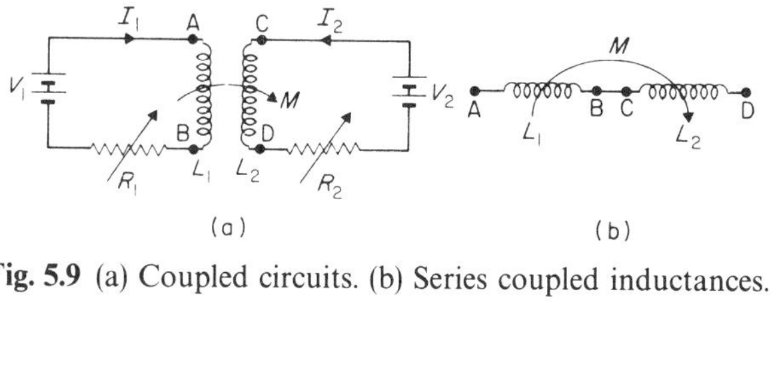

Fig. 5.9 (a) Coupled circuits. (b) Series coupled inductances.

Therefore

$$M_{12}M_{21} = M^2 = N_1 N_2 \Phi_{22} \Phi_{11}/I_1 I_2$$

and, from (5.14) defining self-inductance,

$$M^2 = L_1 L_2$$

or

$$M = \sqrt{L_1 L_2} \tag{5.19}$$

This is the maximum value of M for tight coupling. In general there is some flux leakage and

$$M = k\sqrt{L_1 L_2} \tag{5.20}$$

where $0 \leqslant k \leqslant 1$ and k is the coupling coefficient.

5.2.6 Magnetic energy

A coil with a current flowing in it stores magnetic energy, an outstanding example being a superconducting coil in its persistent mode. In this state the current in the coil flows without any resistance, as long as it is kept below its transition temperature from normal conduction to superconduction. This energy is due to the rate of doing electrical work W by the induced e.m.f. $\mathscr{E}$ in the coil against the induced current I. So,

$$\frac{\mathrm{d}W}{\mathrm{d}t} = \mathscr{E}I = -LI\frac{\mathrm{d}I}{\mathrm{d}t}$$

Therefore, for a perfect (loss-less) coil, the total energy stored is

$$U = -W = \tfrac{1}{2}LI^2 \tag{5.21}$$

6

Magnetism

In this chapter we extend magnetostatics to include matter in general, where previously we have considered the magnetic fields arising from electrons in a vacuum or free electrons in metals. We use Gauss's law for magnetic flux

$$\int_S \boldsymbol{B} \cdot \mathbf{d}\boldsymbol{S} = 0$$

Ampère's circulation law for steady currents

$$\oint_C \boldsymbol{B} \cdot \mathbf{d}\boldsymbol{s} = \mu_0 \int_S \boldsymbol{j} \cdot \mathbf{d}\boldsymbol{S}$$

and we consider induced currents from Faraday's law

$$\oint_C \boldsymbol{E} \cdot \mathbf{d}\boldsymbol{s} = -\int_S \frac{\partial \boldsymbol{B}}{\partial t} \cdot \mathbf{d}\boldsymbol{S}$$

All states of matter will be considered: gases, liquids and solids, especially materials like iron with unusual magnetic properties. We first discuss the microscopic nature of magnetism, then define magnetization $\boldsymbol{M}$, the magnetizing field $\boldsymbol{H}$ and find the boundary conditions for $\boldsymbol{B}$ and $\boldsymbol{H}$ between two media. We continue with an introduction to each of the commonest types of magnetism, dia-, para- and ferromagnetism, and conclude by discussing the production of both permanent and powerful magnetic fields.

6.1 MAGNETIZATION OF MATTER

The simplest model of an atom, the Bohr model (Fig. 6.1(a)) has at least one electron rotating in a fixed, circular orbit about the nucleus.

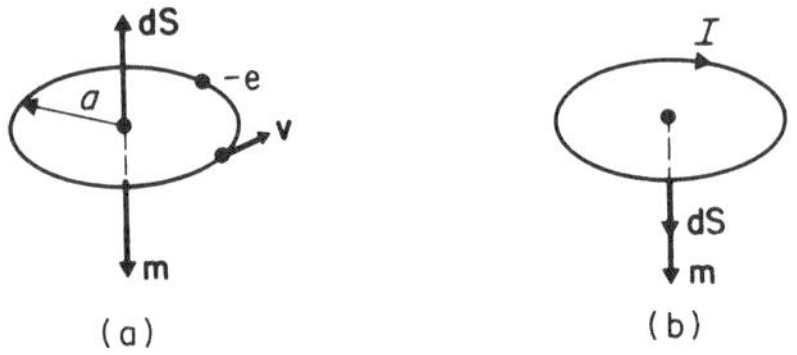

Fig. 6.1 (a) An orbiting electron in an atom is equivalent to (b) a magnetic dipole.

If the orbit has area $dS = \pi a^2$ and the electron is orbiting at a speed v, its motion is equivalent to a current

$$I = -ev/2\pi a$$

From a distance this orbiting electron is equivalent (Fig. 6.1(b)) to a magnetic dipole of moment

$$\boldsymbol{m} = I\mathbf{d}\boldsymbol{S} \tag{6.1}$$

Typical values of $a = 50\,\text{pm}$, $v = 10^6\,\text{m s}^{-1}$ show that m is about $10^{-23}\,\text{A m}^2$. Although this is a very small moment, significant magnetization $\boldsymbol{M}$ of matter can result from the partial or complete alignment of many moments $\boldsymbol{m}$ in a volume V. The total magnetic moment is the vector sum of the individual moments and we define *magnetization* as the total magnetic moment per unit volume,

$$\boldsymbol{M} = \Sigma \boldsymbol{m}/V \tag{6.2}$$

The magnetic moments in matter arise not only from the orbital motion of electrons in atoms (atomic moments) but also from the intrinsic angular momentum (spin) of electrons (intrinsic moments) and from the spin of nuclei (nuclear moments). Magnetization of matter by applied magnetic fields is a similar phenomenon to the polarization of matter by applied electric fields. We found that the polarization $\boldsymbol{P}$ of a dielectric was associated with an induced surface charge density, σ_p. We shall now show that the magnetization $\boldsymbol{M}$ of matter is associated with an induced surface current density, i_m.

In Fig. 6.2(a) an elementary volume $dx\ dy\ dz$ of matter has been magnetized parallel to the applied field $\boldsymbol{B}$ (this is the case for a parallel or *paramagnetic material*, in contrast to an antiparallel or *diamagnetic* matcrial). The atomic magnetic moments add vectorially to give a total moment $\boldsymbol{m}_A$. This can be exactly equivalent to a single current loop (Fig. 6.2(b)) of current I_m around the volume element

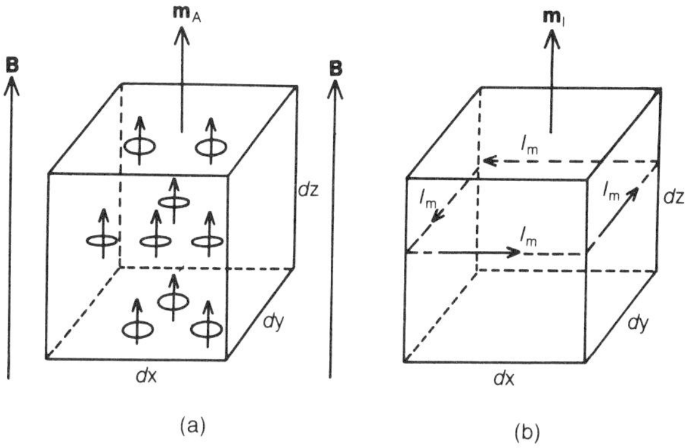

Fig. 6.2 An elementary volume of magnetized matter is equivalent to (b) a surface magnetization current, I_m.

producing a magnetic moment:

$$m_I = I_m \, dx \, dy = m_A$$

There the magnetization of the elementary volume is of magnitude

$$M = \frac{m_A}{dx \, dy \, dz} = \frac{I_m}{dz}$$

and in the same direction as the applied field $\boldsymbol{B}$. The term *magnetization surface current density*, i_m, is used for the surface current per unit length normal to the current, so that here

$$M = \frac{I_m}{dz} = i_m \tag{6.3}$$

which can be compared with $P = \sigma_p$.

In order to relate $\boldsymbol{M}$ to $\boldsymbol{B}$, we first consider a larger volume (Fig. 6.3) of matter uniformly magnetized along $0z$. If this is divided into many elementary volumes $dx \, dy \, dz$, it is clear that the magnetization currents I_m in each small cube will cancel at all the internal surfaces of adjacent cubes and also at all internal edges of the surfaces in the xy planes (see inset of the figure). There remain the surface currents at the other four faces in the xz and yz planes and these all act together to provide a surface current density $i_m = I_m/dz$, as before. Although we have proved this only for a cube, it is easy to see that i_m circulates around $\boldsymbol{M}$ as the current in a long solenoid circulates around $\boldsymbol{B}$ (Figs. 4.16 and 6.4(a)).

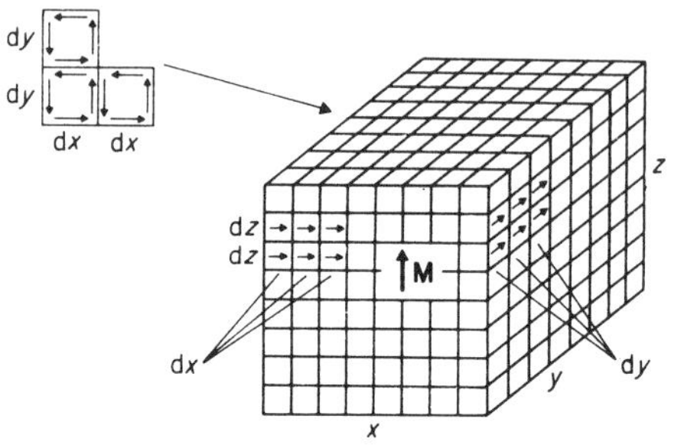

Fig. 6.3 A magnetized volume of matter is equivalent to a surface current density.

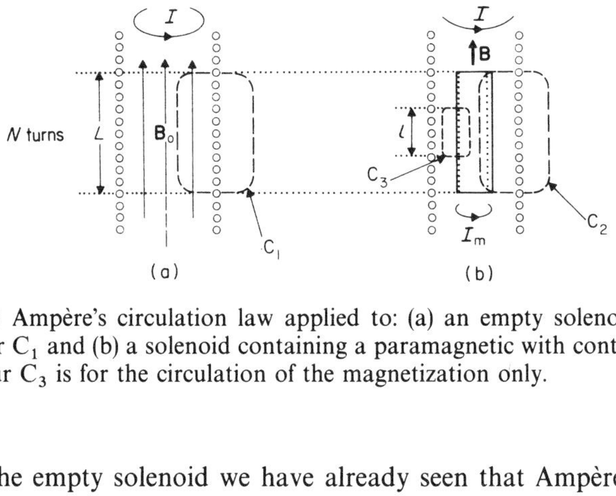

Fig. 6.4 Ampère's circulation law applied to: (a) an empty solenoid with contour C_1 and (b) a solenoid containing a paramagnetic with contour C_2. Contour C_3 is for the circulation of the magnetization only.

In the empty solenoid we have already seen that Ampère's law, for the uniform central field B_0, gives

$$\oint_{C_1} \boldsymbol{B}\cdot\mathbf{d}\boldsymbol{s} = \mu_0 NI = B_0 L$$

This can be expressed as

$$B_0 = \mu_0(NI/L) = \mu_0 i_f \tag{6.4}$$

where i_f is a *solenoidal surface current density* due to the free electrons in the conduction current I with units ampere-turns per metre. When a paramagnetic rod is placed entirely in this central field, it is magnetized and its magnetization M is equivalent to a uniform surface current density i_m. Clearly these surface currents increase the field in the solenoid. According to Ampère's law

$$\oint_{C_2} \boldsymbol{B}\cdot\mathbf{d}\boldsymbol{s} = \mu_0(NI + i_m L) = BL$$

or

$$B = \mu_0(i_f + i_m) \tag{6.5}$$

We have already shown that the magnetization M is equal to i_m and, for a paramagnetic, is in the same direction as B. We now define the *magnetizing field* $\boldsymbol{H}$ as that due to the solenoidal current density alone, so that $H = i_f$. For the empty solenoid, (6.4) can be rewritten

$$\boldsymbol{B}_0 = \mu_0 \boldsymbol{H} \tag{6.6}$$

and this is only true when there is no matter present. Thus $\boldsymbol{H}$ like $\boldsymbol{D}$ ignores the effects of introducing matter and is entirely due to the free, conduction electrons of magnetostatics. When matter is present, (6.5) can therefore be written

$$\boldsymbol{B} = \mu_0(\boldsymbol{H} + \boldsymbol{M}) \tag{6.7}$$

When the magnetic matter is linear, isotropic and homogeneous, $\boldsymbol{M}$ is proportional to $\boldsymbol{H}$ and we define the *magnetic susceptibility*, χ_m, as the dimensionless ratio M/H. This is the definition commonly accepted, but some authors have used $\chi_B = \mu_0 M/B$, which is also dimensionless. The *magnetic permeability* is defined as $\boldsymbol{B}/\boldsymbol{H}$, so that (6.7) can be written

$$\mu = \frac{B}{H} = \mu_0(1 + \chi_m) \tag{6.8}$$

or alternatively as

$$\mu = \frac{B}{H} = \frac{\mu_0}{(1 - \chi_B)}$$

For paramagnetics in which $\chi_m \ll 1$ and positive, or diamagnetics in which $\chi_m \ll 1$ and negative, there is little difference between χ_B and χ_m. But in ferromagnetics where $\chi_m \gg 1$ and positive, M is only proportional to H for small ranges of H and χ_B is not the same as χ_m. The relative permeability $\mu_r = \mu/\mu_0$ is given by

$$\mu_r = 1 + \chi_m \tag{6.9}$$

6.2 CHARACTERISTICS OF *B* AND *H*

6.2.1 Circulation

We have seen that the circulation of ***B***, given by Ampère's law

$$\oint_C \boldsymbol{B}\cdot\mathbf{d}\boldsymbol{s} = \mu_0 I$$

can be written in the presence of magnetized matter in a solenoid as

$$\oint_C \boldsymbol{B}\cdot\mathbf{d}\boldsymbol{s} = \mu_0(I_f + I_m)$$

where I_f is the free, conduction current and I_m is the magnetization current. Using (6.7) this becomes

$$\oint_C \boldsymbol{H}\cdot\mathbf{d}\boldsymbol{s} + \oint_C \boldsymbol{M}\cdot\mathbf{d}\boldsymbol{s} = I_f + I_m$$

From Fig. 6.4(b) the contour C_3 shows that the circulation of ***M*** is just I_m, since the contributions to $\boldsymbol{M}\cdot\mathbf{d}\boldsymbol{s}$ are $i_m l$ within the paramagnetic, zero normal to ***M*** and zero outside the paramagnetic.

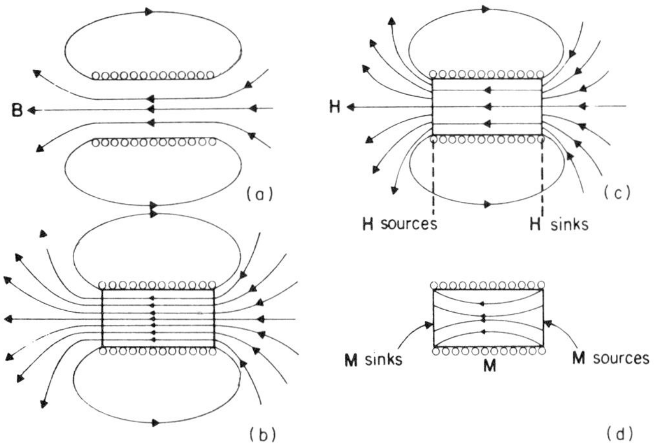

Fig. 6.5 Magnetic field ***B*** for (a) an empty solenoid and (b) a solenoid filled with a paramagnetic of relative permeability μ_r for which the magnetizing field ***H*** lines are in (c) and the magnetization ***M*** lines are in (d).

Therefore, by subtraction, the circulation of $\boldsymbol{H}$ is

$$\oint_C \boldsymbol{H}\cdot\mathbf{d}\boldsymbol{s} = I_{\mathrm{f}} \tag{6.10}$$

and this shows the importance of $\boldsymbol{H}$ as the magnetizing field from conduction currents. However, it is only $\boldsymbol{B}$ that satisfies Gauss's law, $\int_S \boldsymbol{B}\cdot\mathbf{d}\boldsymbol{S} = 0$, and therefore only the lines of $\boldsymbol{B}$ are continuous with no sources or sinks (Fig. 6.5(a), (b)). At the surface of magnetized matter the lines of $\boldsymbol{M}$ disappear (Fig. 6.5(d)), while the corresponding lines of $\boldsymbol{H}$ begin there (Fig. 6.5(c)). This figure also shows how a magnetic medium of relative permeability μ_{r} increases $\boldsymbol{B}(=\mu_{\mathrm{r}}\mu_0\boldsymbol{H})$, in contrast to a dielectric of relative permittivity ε_{r} which decreases $\boldsymbol{E}(=\boldsymbol{D}/\varepsilon_{\mathrm{r}}\varepsilon_0)$.

6.2.2 Boundary relations

How does the magnetic field change when it crosses the boundary between two magnetic media of permeabilities $\mu_1 = \mu_{\mathrm{r}1}\mu_0$ and $\mu_2 = \mu_{\mathrm{r}2}\mu_0$? To find out we apply Gauss's law for $\boldsymbol{B}$ and the circulation law for $\boldsymbol{H}$ to the magnetic vectors shown in Fig. 6.6. Our treatment parallels that for $\boldsymbol{D}$ and $\boldsymbol{E}$.

For the flux into and out of the Gaussian cylinder of cross-section dS and negligible height we have

$$\boldsymbol{B}_1\cdot\mathbf{d}\boldsymbol{S}_1 + \boldsymbol{B}_2\cdot\mathbf{d}\boldsymbol{S}_2 = 0$$

Hence only the normal component B_{n} of each magnetic field

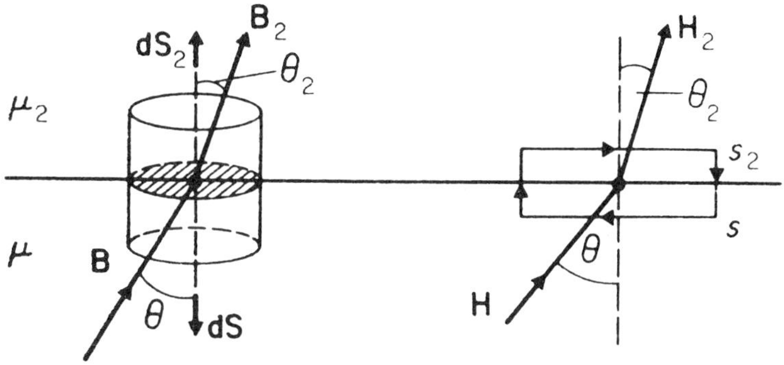

Fig. 6.6 Boundary conditions for the magnetic vectors **B** and **H** crossing between two magnetic media of permeabilities μ_1 and μ_2.

contributes and

$$B_{1n} = B_{2n} \tag{6.11}$$

Applying the circulation law to the magnetizing fields $\boldsymbol{H}_1$ and $\boldsymbol{H}_2$ crossing the closed loop of length s_1 and s_2, we have

$$\int_C \boldsymbol{H} \cdot \mathbf{d}\boldsymbol{s} = \boldsymbol{H}_1 \cdot s_1 + \boldsymbol{H}_2 \cdot s_2 = 0$$

since there are no conduction currents present at the boundary of two magnetic media (except superconductors). The loop can be as near as we wish to the surface and so only the tangential component H_t of each magnetizing field contributes and

$$H_{1t} = H_{2t} \tag{6.12}$$

At the boundary we therefore have continuity for B_n and H_t and when it is valid to write $\boldsymbol{B} = \mu_r \mu_0 \boldsymbol{H}$, we can write (6.11) and (6.12) as

$$\mu_{r1} H_1 \cos\theta_1 = \mu_{r2} H_2 \cos\theta_2$$

$$H_1 \sin\theta_1 = H_2 \sin\theta_2$$

We therefore get refraction of $\boldsymbol{B}$ and $\boldsymbol{H}$ at the boundary with the relation

$$\frac{\tan\theta_1}{\tan\theta_2} = \frac{\mu_{r1}}{\mu_{r2}} \tag{6.13}$$

6.2.3 Energy density

For a coil of self-inductance L we showed that the energy stored in it (5.21), is

$$U = \tfrac{1}{2} L I^2$$

where, from (5.17), $L = \mu_0 N^2 S / l$. Applying these equations to the empty solenoid of Fig. 6.4 we have

$$B_0 = \mu_0 N I / l$$

and

$$\boldsymbol{B}_0 = \mu_0 \boldsymbol{H}$$

Therefore,

$$U = \frac{\mu_0}{2}\left(\frac{N^2 S I^2}{l}\right)$$

and the energy density

$$u = \frac{\mu_0}{2}\left(\frac{N^2 I^2}{l^2}\right) = \frac{1}{2}\boldsymbol{B}_0 \cdot \boldsymbol{H}$$

since $\boldsymbol{B}_0$ and $\boldsymbol{H}$ are parallel inside a long solenoid.

For a filled solenoid, if the magnetic material is linear, isotropic and homogeneous then $\boldsymbol{B} = \mu \boldsymbol{H}$ (6.8), and the inductance increases to $L = \mu N^2 S/l$. Therefore the energy density is

$$u = \frac{\mu}{2}\left(\frac{N^2 I^2}{l^2}\right) = \frac{1}{2}\boldsymbol{B} \cdot \boldsymbol{H} \tag{6.14}$$

We showed in section 2.4 that electrostatic energy is stored in the electric field and a similar argument can be used to show that magnetic energy is stored in the magnetic field. Therefore, we can write for the total magnetic energy

$$U = \frac{1}{2}\int_\tau \boldsymbol{B} \cdot \boldsymbol{H} \,\mathrm{d}\tau \tag{6.15}$$

which is the corresponding equation to (2.36) for electric energy.

6.3 MAGNETISM IN MATTER

For most substances magnetic effects are very small and can only be seen using high magnetic fields or sensitive detectors like those involved in nuclear magnetic resonance experiments or magnetic resonance imaging (MRI) in hospitals. In these substances the relative permeability $\mu_r \ll 1$, but in ferromagnetics such as iron, cobalt and nickel, $\mu_r \gg 1$ and is found from $\mathrm{d}B/(\mu_0 \mathrm{d}H)$, as $\boldsymbol{B}$ is not proportional to $\boldsymbol{H}$. Here we consider first diamagnetics, in which the induced magnetism acquires a dipole moment opposed to the applied field, and then paramagnetism where permanent dipole moments are aligned parallel to the applied field. In each we consider only the orbital motion of electrons in atoms, although most atomic magnetism arises from the intrinsic moments of the electrons associated with their spin. However, the principles of dia- and paramagnetism can be understood in terms of the orbital moments.

6.3.1 Diamagnetics

An atomic dipole arises from an orbiting electron on Bohr theory (Figs. 6.1 and 6.7) and has moment, from (6.1),

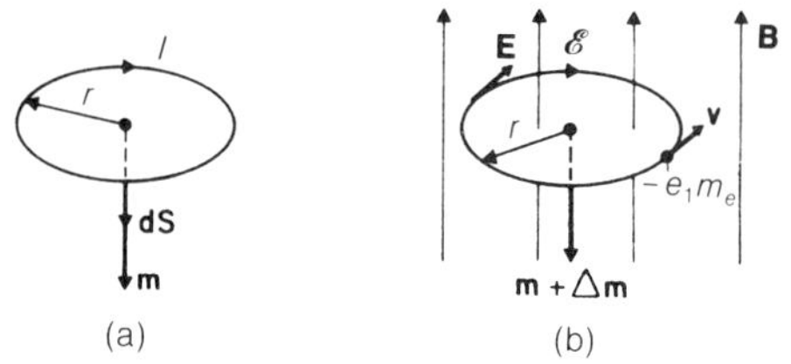

Fig. 6.7 (a) An atomic dipole of moment $\boldsymbol{m}$. (b) Enhanced to $\boldsymbol{m} + \Delta\boldsymbol{m}$ by the applied field $\boldsymbol{B}$.

$$\boldsymbol{m} = \boldsymbol{I}\,\mathrm{d}\boldsymbol{S} = \frac{-ev}{2\pi r}(\pi r^2)\hat{\mathbf{n}} = \frac{-evr}{2}\hat{\mathbf{n}} \tag{6.16}$$

where $\hat{\mathbf{n}}$ is the normal for the electron orbit in Fig. 6.7(b). When a uniform magnetic field is applied to this electron it induces an e.m.f. $\mathscr{E}$ in the 'circuit' given by Faraday's law (5.8)

$$\mathscr{E} = \oint_C \boldsymbol{E}\cdot\mathbf{ds} = -\int_S \frac{\partial B}{\partial t}\,\mathbf{d}\boldsymbol{S}$$

which becomes

$$2\pi \mathrm{r} E = -\frac{\mathrm{d}B}{\mathrm{d}t}(\pi r^2)$$

By Newton's second law, the electron of mass m_e is accelerated by the force $-e\boldsymbol{E}$ acting on it. Thus,

$$m_e\frac{\mathrm{d}v}{\mathrm{d}t} = -eE = \frac{er}{2}\frac{\mathrm{d}B}{\mathrm{d}t}$$

Integrating this over the time the field increases from 0 to B and the speed v increases by Δv, where

$$\Delta v = \left(\frac{er}{2m_e}\right)B \tag{6.17}$$

producing an incremental moment from (6.16) of

$$\Delta m = -\frac{er}{2}\Delta v = -\frac{e^2r^2}{4m_e}B$$

which by Lenz's law opposes $\boldsymbol{B}$ so that

$$\Delta\boldsymbol{m} = \frac{-e^2r^2}{4m_e}\boldsymbol{B} \tag{6.18}$$

The value of the orbital radius r results from a balance between the Coulomb attraction $\boldsymbol{F}$ of the nucleus for the electron and the centrifugal force $m_e v^2/r$. The Lorentz force due to $\boldsymbol{B}$ on the electron is $-e\boldsymbol{v}\times\boldsymbol{B}$ and this acts to increase F by evB. On the other hand, the increased speed Δv produces an increase $(2m_e v\Delta v/r + m_e(\Delta v)^2/r)$ in the centrifugal force and this, from (6.17), balances the increase in F if $\Delta v \ll 2v$. Simple calculations give $\Delta v \sim 100\,\mathrm{m\,s^{-1}}$ and $v \sim 10^6\,\mathrm{m\,s^{-1}}$, so that r is constant. Therefore the effect of $\boldsymbol{B}$ is to produce a larger negative moment $(\boldsymbol{m}+\Delta\boldsymbol{m})$, where $\Delta\boldsymbol{m}$ (negative) is given by (6.18), but to maintain the electron in its original orbit.

The magnetization of a diamagnetic containing N atoms per unit volume, each with Z orbiting electrons on radii with a mean square radius $\langle r^2\rangle$ is therefore

$$\boldsymbol{M} = N\Delta\boldsymbol{m} = \left(\frac{-e^2NZ\langle r^2\rangle}{4m_e}\right)\boldsymbol{B}$$

and the corresponding susceptibility is

$$\chi_B = \frac{\mu_0 M}{B} = -\frac{\mu_0 e^2 N}{4m_e} Z\langle r^2\rangle \simeq \chi_m \tag{6.19}$$

One possible model for such a diamagnetic is that its atomic electrons are paired off with opposite rotations and so opposing dipole moments (Fig. 6.8(a)). Application of a magnetic field $\boldsymbol{B}$ then accelerates the electron in orbit 1 and slows down that in orbit 2 with the net result that they acquire a moment $2\Delta\boldsymbol{m}$ opposing $\boldsymbol{B}$ (Fig. 6.8(b)). In terms of the angular velocity $\omega = v/r$ of each electron,

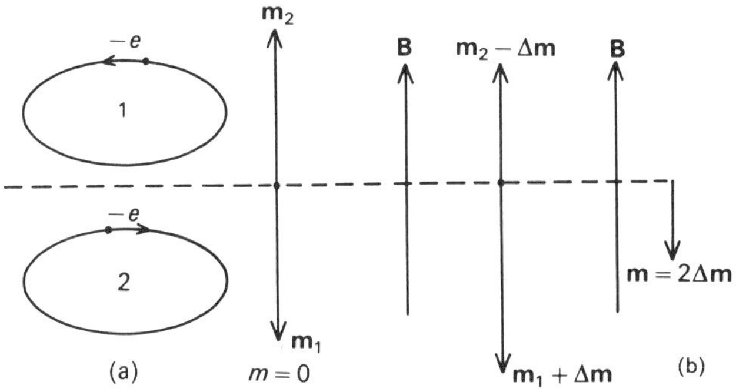

Fig. 6.8 A diamagnetic could consist of an electron pair with opposing dipole moments. (a) $\boldsymbol{B}=0$. (b) $\boldsymbol{B}$ produces a moment $2\Delta\boldsymbol{m}$.

the change in angular velocity is called the *Larmor (angular) frequency*, ω_L, which from (6.17), is

$$\omega_L = eB/2m_e \tag{6.20}$$

The angular velocity of orbit 1 thus becomes $(\omega_0 + \omega_L)$ and that of orbit 2, $(\omega_0 - \omega_L)$.

This model of a diamagnetic implies that χ_m is independent of temperature, since the core structure of an atom is normally unchanged by thermal vibrations. For a mole of solid diamagnetic occupying $10\,\text{cm}^3$, $N = 6 \times 10^{28}\,\text{m}^{-3}$ and if we put $Z = 29$, $\langle r^2 \rangle = a_0^2$, (6.19) predicts χ_m of about -40×10^{-6}, which is a similar order of magnitude to χ_m measured for copper (-10×10^{-6}) and glass (-110×10^{-6}). Similar for a gas at STP with $N = 6 \times 10^{25}\,\text{m}^{-3}$, χ_m is about -40×10^{-9}, which is not very different from χ_m for nitrogen gas (-5×10^{-9}).

6.3.2 Paramagnetics

In paramagnetics each atom has a small permanent dipole moment $\boldsymbol{\mu}$ due to its orbiting electrons (and their spin). On Bohr theory each electron with $\mu = \frac{1}{2}evr$ (6.16) has angular momentum $m_e vr = n\hbar$, where n is the number of the orbit and so

$$\mu = n(e\hbar/2m_e) \tag{6.21}$$

The moment of an electron in the first Bohr orbit is called the *Bohr magneton* μ_B, and is a unit of about $9 \times 10^{-24}\,\text{A}\,\text{m}^2$. When the moments of the core electrons in a paramagnetic atom or molecule have been added vectorially, the total moment should be of the order of $10^{-23}\,\text{A}\,\text{m}^{-2}$, if other effects are not present.

How many of these dipoles are likely to be aligned when we apply a magnetic field? If we do the experiment at room temperature the thermal energy of about $k_B T = 1.4 \times 10^{-23} \times 300$ joules, roughly 25 meV, will tend to keep them pointing in random directions. We saw in Chapter 4 that there is a torque $\boldsymbol{m} \times \boldsymbol{B}$ (4.13) tending to align a magnetic dipole with an applied field. An unaligned dipole therefore has a greater potential energy than an aligned one. If we take the zero of potential energy, which is arbitrary, as that when $\boldsymbol{m}$ is normal to $\boldsymbol{B}$, then the energy lost by the dipole during alignment is $\boldsymbol{m}\cdot\boldsymbol{B}$. Therefore the potential energy of the dipole in the field is

$$U = -\boldsymbol{m}\cdot\boldsymbol{B} \tag{6.22}$$

and the maximum energy that could be given to a dipole would be to completely reverse its polarity from $-\boldsymbol{m}$ to $+\boldsymbol{m} = 2mB$. If m is about to 10^{-23} A m^2 and a large field, say 5 T, is applied, U will not exceed 10^{-22} joules or roughly 1 me V.

Thus only a small fraction f of the atomic dipoles $\boldsymbol{\mu}$ will be aligned parallel to $\boldsymbol{B}$ and it can be shown, from statistical mechanics, that this fraction is

$$f = \frac{\mu B}{3kT} \tag{6.23}$$

Typically $\mu B \sim 1$ meV, while $3k_B T \sim 75$ meV, so that $f \simeq 1\%$. Hence the magnetization of a paramagnetic with N atoms per unit volume is, from (6.2),

$$\boldsymbol{M} = \frac{\Sigma\mu}{V} = N\mu f$$

or

$$\boldsymbol{M} = \left(\frac{N\mu^2}{3k_B T}\right)\boldsymbol{B}$$

and

$$\chi_m = \frac{M}{H} = \frac{\mu_0 N \mu^2}{3k_B T} \tag{6.24}$$

This is known as *Curie's law* and shows that χ_m increases as $1/T$ at low temperatures, in contrast to the diamagnetics, where χ_m in (6.19) is independent of temperature.

Some paramagnetic salts, such as chromium potassium alum, can be magnetically saturated at low temperatures with a superconducting solenoid providing, say 2 T at 1 K. Others, such as cerium magnesium nitrate, continue to obey Curie's law to very low (millikelvin) temperatures and so their reciprocal susceptibility as a function of temperature provides an excellent thermometer. These are in direct contrast to electric dipole moments, which usually cannot be saturated at low temperatures, since they rely on the displacement of electron distributions (in atoms) or their rotation (in molecules), movements that would be frozen in most solids.

In metals the major contribution is the Pauli spin paramagnetism of the conduction electrons, in which at a temperature T only the small fraction T/T_F, where T_F is the Fermi temperature of about 10^5 K, take part. This is discussed in Professor Chambers' book in this series.

In gases the total susceptibility on the Bohr model presented here would be

$$\chi_m = N\mu_0\left(\frac{\mu^2}{3k_B T} - \frac{e^2 Z\langle r^2\rangle}{4m_e}\right) \tag{6.25}$$

and experiments over a range of temperatures enable both μ and $\langle r^2\rangle$ to be measured. The moments found are of the order of a Bohr magneton in a paramagnetic gas like oxygen and in the diamagnetic inert gases the radii of the atoms are about one Bohr radius (5×10^{-11} m), so that Bohr's theory does give reasonable values for the static magnetic susceptibility of gases.

6.3.3 Ferromagnetics

Ferromagnetics are materials in which magnetization occurs spontaneously, that is without the application of an external magnetic field. The commonest ferromagnetic is iron and its alloys (steels), but it is well known that an iron rod is not a good magnet. There are two reasons for this, shown in Figs. 6.9 and 6.10. First, the

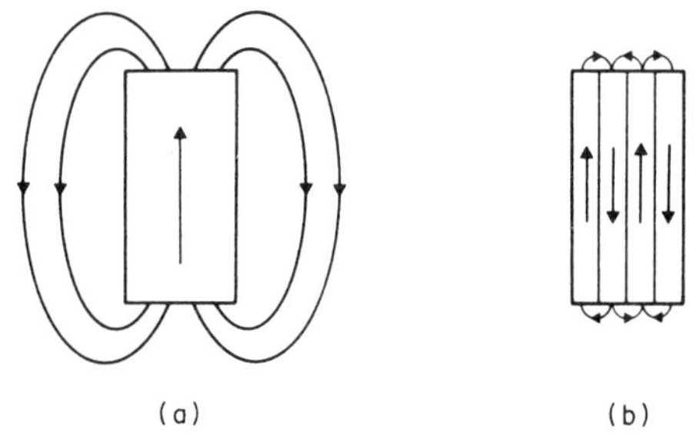

Fig. 6.9 (a) A single domain and (b) four domains in a single crystal of a ferromagnetic.

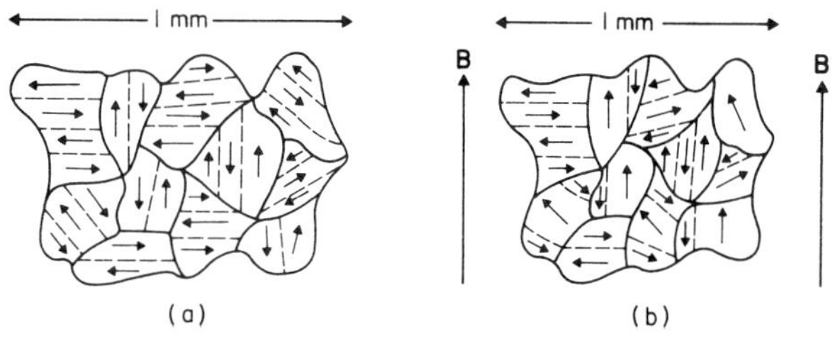

Fig. 6.10 A ferromagnetic (a) in its unmagnetized state, and (b) in a magnetic field.

material divides up into regions which are perfect magnets, known as *domains*, but these domains are magnetized in many different directions. In Fig. 6.9 a ferromagnetic which consisted of only one domain is compared with one having four domains, magnetized alternatively in opposite directions. The single domain has a large external magnetic field, which is energetically less favourable than the multidomain sample with a negligible external field. The size of the domains in practice is a balance between the energy saved through their existence and the energy used in creating the boundaries between them.

The second reason is the existence of many differently orientated small crystals, known as *grains*, in ordinary, polycrystalline iron. The net result is that spontaneously magnetized iron at room temperature has, under a microscope with a suitable coating, the appearance of Fig. 6.10(a). Each grain consists of several domains, magnetized in a direction corresponding to the crystallographic orientations of the grains. When a magnetic field is applied (Fig. 6.10(b)) the initial effect is for the domains that are favourably orientated to grow in size at the expense of their less fortunate neighbours. Such an effect is reversible when the field is removed and corresponds to the initial magnetizing region, OA in Fig. 6.11(a). A larger magnetic field will start to rotate the domains within the grains, producing greater magnetization, as AB in Fig. 6.11(a). Finally, at sufficiently high fields the magnetization saturates, when all the domains are aligned along the applied field, as at C in Fig. 6.11(a).

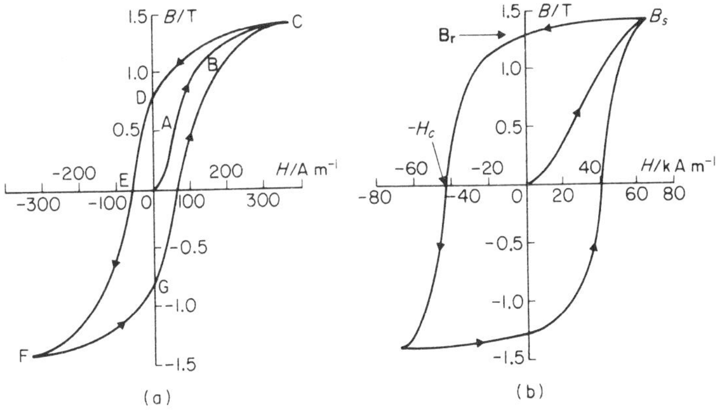

Fig. 6.11 Hysteresis curves (a) for polycrystalline iron, and (b) for a permanent magnet alloy.

Experimentally a B–H curve of this type can be obtained by winding a toroidal coil (Fig. 4.15) around an iron ring, when the magnetizing field is given by $H = NI/l$, where the toroid has N turns over a length l and carries a current I, (6.4) and (6.6), while B is measured by induction in a second coil attached to a fluxmeter (a moving coil meter with negligible restoring torque). Following initial magnetization to saturation, when the current I and so field H are reduced to zero, there is a *remanence* (OD in Fig. 6.11(a)) and for complete demagnetization it is necessary to apply a reversed field OE, known as the *coercivity*. As the field continues to be reversed the magnetization eventually saturates in the opposite direction at F and the cycle is completed by FGC. The complete curve is a *hysteresis loop* and represents a significant loss of energy through its irreversibility.

The loss occurs in the movements of the domain boundaries and comes from the source of the current in the coil. By Faraday's law an e.m.f. $\mathscr{E}$ is induced in the winding as B changes given by

$$\mathscr{E} = -NS\frac{\mathrm{d}B}{\mathrm{d}t}$$

where S is the cross-sectional area of the iron ring. This e.m.f. will oppose the increase in the current and so the extra power from the source to maintain the current will be

$$\frac{\mathrm{d}U}{\mathrm{d}t} = -I\mathscr{E} = INS\frac{\mathrm{d}B}{\mathrm{d}t} = HSl\frac{\mathrm{d}B}{\mathrm{d}t}$$

The volume of the ring is Sl and so the energy density in a complete hysteresis loop will be

$$u = \oint H\,\mathrm{d}B \qquad (6.26)$$

where the integral is just the area of the loop on the B–H plot.

Materials such as soft (pure) iron have comparatively small hysteresis loops (Fig. 6.11(a)) and so are used for alternating currents in chokes (self-inductances) and transformers, while specially selected iron alloys (steels) can have very large loops (Fig. 6.11(b)). Here the saturation field B_S is similar to that for iron, but the magnetizing and demagnetizing fields are very much larger (note the change of scale in H from $\mathrm{A\,m^{-1}}$ to $\mathrm{kA\,m^{-1}}$). The result is that the area of the loop and the coercivity $(-H_c)$ are much larger and so these alloys make excellent permanent magnets. Typically the energy density in a hysteresis loop ranges from about $100\,\mathrm{J\,m^{-3}}$ in a soft ferromagnetic to about $10^5\,\mathrm{J\,m^{-3}}$ in a very hard one.

When ferromagnetics are heated above their *Curie point*. T_C, they become paramagnetics obeying a Curie–Weiss law,

$$\chi_m = \frac{C}{T - T_C} \tag{6.27}$$

which differs from Curie's law (6.24). At the Curie point the interactions between the electron spins undergo a cooperative transition, known as an order–disorder transition, which results in the spontaneous magnetization and the formation of the domains. In Fig. 6.12 the resulting magnetization is seen to increase sharply just below T_C and then to saturate at low temperatures. It is only in recent years that theories in quantum statistics have been devised to explain such very strong interactions.

A special class of *ferrimagnetics* are ferromagnetic insulators, such as ferrites (oxides of iron), which have almost rectangular hysteresis loops and so can be switched rapidly from $-B_s$ to $+B_s$. This makes them particularly suitable for the cores of high frequency transformers, for magnetic tapes and for computer memories. In a few materials called *antiferromagnetics*, such as chromium, magnetic interactions between the atomic spins cause the spins to be alternatively pointing in opposite directions and so the net magnetization is quite small. However they do exhibit spontaneous magnetization below a temperature called the *Néel point* and show similar nonlinear behaviour to ferromagnetics.

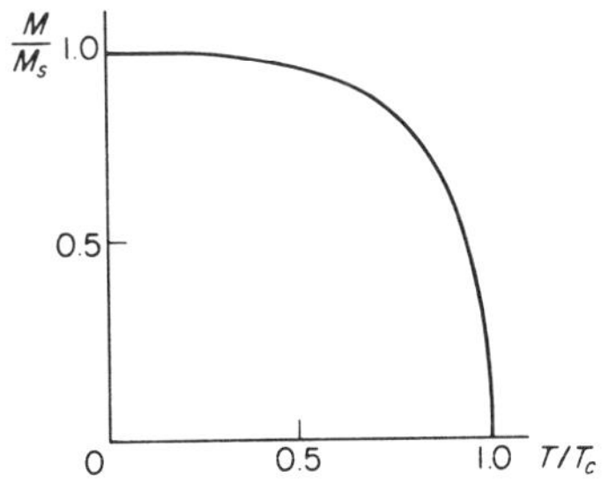

Fig. 6.12 Spontaneous magnetization of a ferromagnetic when cooled below its Curie point, T_c.

6.4 PRODUCTION OF MAGNETIC FIELDS

There are three basic arrangements for producing strong magnetic fields: solenoids, electromagnets and permanent magnets. We consider each of these in turn and apply the following equations to solve the magnetic circuits involved:
Gauss's law

$$\int_S \boldsymbol{B}\cdot\mathbf{d}\boldsymbol{S}=0 \tag{4.28}$$

Circulation law for $\boldsymbol{H}$

$$\oint_C \boldsymbol{H}\cdot\mathbf{d}\boldsymbol{s}=I_f \tag{6.10}$$

For magnetic media

$$\boldsymbol{B}=\mu_r\mu_0\boldsymbol{H} \tag{6.8}$$

can be applied at a point on the hysteresis loop as

$$\mu_r = \mathrm{d}B/(\mu_0 \mathrm{d}H)$$

6.4.1 Solenoids

These rely on a high current to provide an axial field $\mu_0 NI/l$ (Fig. 6.4(a)). The highest steady fields are provided by Bitter solenoids formed from stacks of water-cooled, perforated copper discs having up to 1000 A at 200 V from powerful generators passing through them. For a bore of 3–4 cm fields of 25–20 T are available in several countries, but such installations are extremely expensive to build and operate. For most purposes superconducting solenoids, using fine wires of niobium–titanium (Nb–Ti) or niobium–tin (Nb_3Sn) embedded in a copper matrix, are used to produce fields in the range 1–15 T from laboratory sizes to large installations for the confinement of plasmas or hydrogen bubble chambers. They operate in liquid helium at 4.2 K, but once the currents have been established in them they persist indefinitely with negligible electrical power loss. Laboratory magnets can have room temperature axial holes, for example, for optical or magnetic resonance experiments.

The surprising discovery in 1986 by Bednorz and Müllcr that an oxide (barium–lanthanum–copper oxide) became superconducting at the 'high' temperature of 30 K, for which they were awarded the

Nobel Prize for Physics in 1987, led to a major search for other oxides that would become superconducting at still higher temperatures. Significantly yttrium–barium–copper oxide was found to be superconducting when cooled below 90 K, using liquid nitrogen or liquid air. This discovery can be expected to lead to the development of superconducting motors and high field magnets operating in liquid nitrogen at 77 K, which will be a major advance on such magnets in liquid helium at 4.2 K.

Still higher fields are possible as short pulses due to extremely high discharge currents from banks of low-loss capacitors into small, rigid, single turn coils. Transient fields of 100 T or more have been used for example in magneto-optical spectroscopy. The highest transient fields have been produced by flux compression in magnetic implosions for about a microsecond.

6.4.2 Electromagnets

A laboratory electromagnet consists of an insulated copper coil around a ferromagnetic core to concentrate the flux in a small air-gap and so provide a strong field at room temperature for susceptibility measurements or resonance experiments. It can be analysed as a torus with a core (Fig. 6.13(a)). If we assume negligible flux leakage, (6.10) and (6.8) become

$$\sum \frac{B_i l_i}{\mu_r \mu_0} = NI \tag{6.28}$$

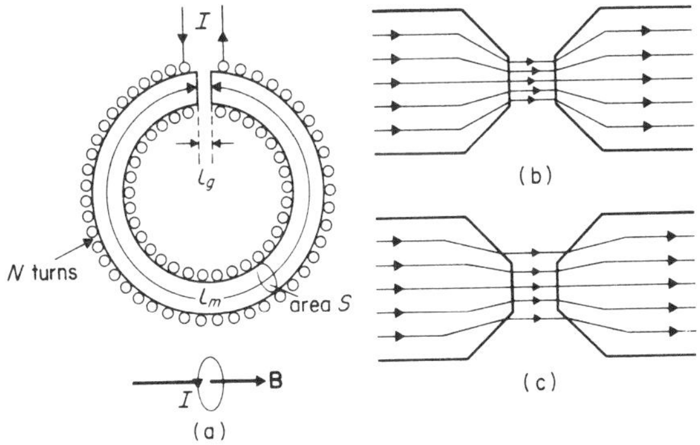

Fig. 6.13 (a) Electromagnet with air gap. (b) Magnetic circuit at the air gap. (c) Flux leakage at the air gap.

where i refers to either the core (m) or the gap (g). The flux Φ is continuous (Fig. 6.13(b)) and so by analogy with Ohm's law, we can write,

$$NI = \Phi \sum \frac{l_i}{\mu_r \mu_0 S_i} = \Phi \sum \mathscr{R} \tag{6.29}$$

where NI is the *magnetomotive force* $\mathscr{F}$ and $\mathscr{R}$ is the magnetic *reluctance* of the circuit. From this equation it follows that reluctances add in series and parallel like resistors and so it does have an application in more complex magnetic circuits. However it ignores the small flux leakage that occurs in practice and is exaggerated in Fig. 6.13(c).

From Fig. 6.13(a) we have

$$\mathscr{R} = \frac{l_m}{\mu_r \mu_0 S} + \frac{l_g}{\mu_0 S} \tag{6.30}$$

and we see that most of the reluctance is in the air gap, since μ_r for a typical core is several thousand. Using (6.29), the flux is

$$\Phi = \frac{\mu_0 NIS}{\left(\dfrac{l_m}{\mu_r} + l_g\right)}$$

and for plane pole pieces

$$B_m = B_g = \frac{\mu_0 NI}{\left(\dfrac{l_m}{\mu_r} + l_g\right)}$$

On the other hand the magnetizing fields are

$$H_g = \frac{NI}{\left(\dfrac{l_m}{\mu_r} + l_g\right)} \quad \text{and} \quad H_m = \frac{NI}{(l_m + \mu_r l_g)}$$

so that the gap in the ring: (1) reduces B drastically; and (2) concentrates H in the gap ($H_g \gg H_m$).

Weiss electromagnets (Fig. 6.14) have large, steerable yokes and are often mounted on rails so that they can be used in more than one experiment. The coils are watercooled, wound on adjustable, soft-iron pole pieces (to give a variable gap) and mounted on a steel yoke. They are normally supplied from stabilized power supplies to

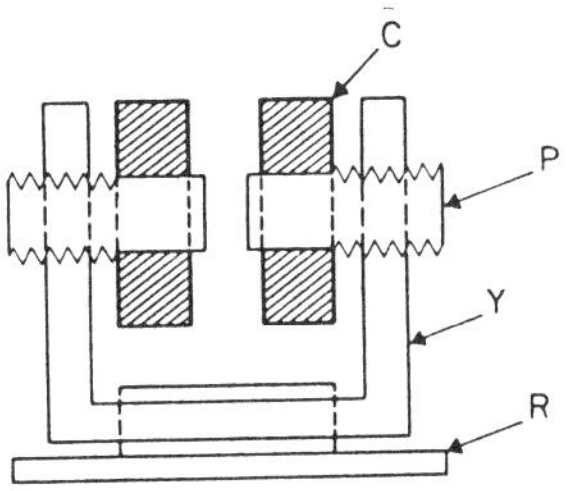

Fig. 6.14 A Weiss electromagnet with coils C mounted on adjustable pole pieces P and a yoke Y on a rotating base R.

give either a uniform field (with flat pole pieces up to 300 mm diameter) or a concentrated field (with tapered pole pieces). The largest Weiss magnet ever built produced about 7 T from 100 kW and weighed 35 tonnes.

6.4.3 Permanent magnets

Permanent magnets are limited by the remanence B_r of known alloys to about 1 T (Fig. 6.15), but are compact, cheap and portable. They are commonly used with magnetrons to generate microwave power in radar, for the non-destructive testing of materials and as standards for calibrating magnetometers.

Applying (6.10) to the magnetic circuit in the permanent magnet

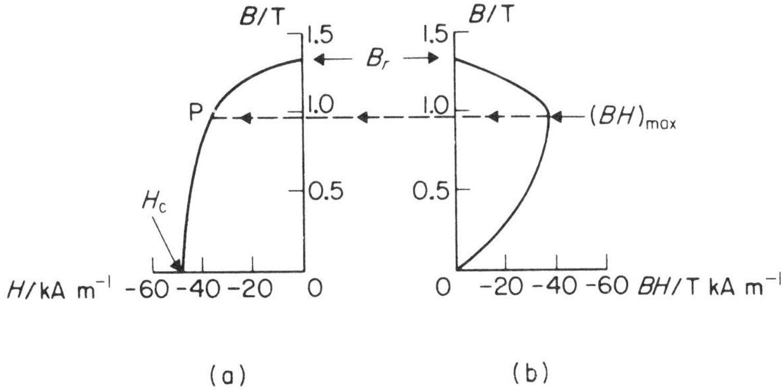

Fig. 6.15 (a) Demagnetization *B–H* curve for a permanent magnetic alloy. (b) *B–BH* curve to find optimum size for a permanent magnet.

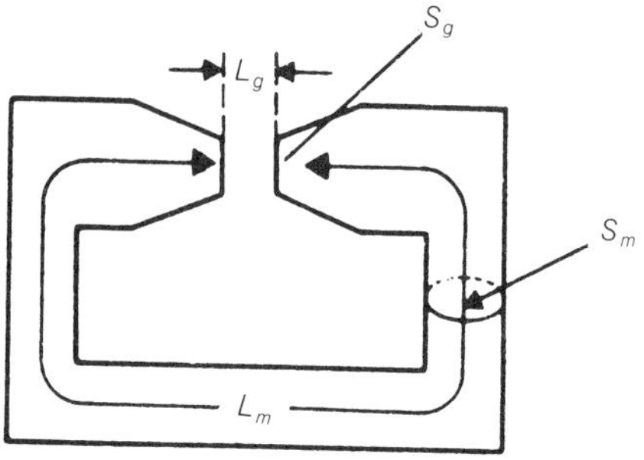

Fig. 6.16 A permanent magnet.

of Fig. 6.16,

$$\oint_C \boldsymbol{H}\cdot d\boldsymbol{s} = H_g l_g + H_m l_m = 0 \qquad (6.31)$$

since there are no conduction currents. Thus H_g and H_m are in opposite directions, so that H_m is the *demagnetizing field* of the magnet. For negligible flux leakage,

$$\Phi = B_g S_g = B_m S_m \qquad (6.32)$$

Multiplying (6.31) and (6.32)

$$B_g H_g l_g S_g = -B_m H_m l_m S_m$$

It is of interest to compare the magnetization of a permanent bar magnet (Fig. 6.17) with that of a paramagnetic material (Fig. 6.5). In

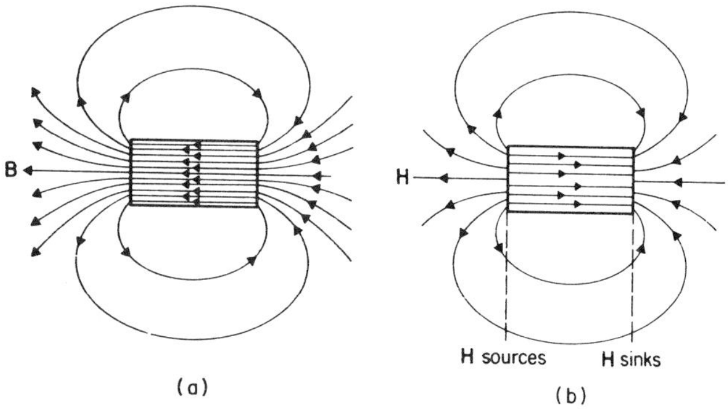

Fig. 6.17 (a) Magnetic field $\boldsymbol{B}$ for a permanent bar magnet. (b) Magnetizing field $\boldsymbol{H}$ outside and demagnetizing field $\boldsymbol{H}_m$ inside a permanent magnet.

both cases the lines of $\boldsymbol{B}$ are continuous and show how $\boldsymbol{B}$ is enhanced in the material. On the other hand the magnetizing field $\boldsymbol{H}$ in Fig. 6.5(c) is replaced by the demagnetizing field $\boldsymbol{H}_m$ in Fig. 6.17(b).

A permanent magnet designed to produce a certain field B_g $(=\mu_0 H_g)$ in a gap of length l_g and cross-section S_g will be of minimum size and cost if it is operated with $(B_m H_m)$ at its maximum value and so minimum volume $l_m S_m$ of the yoke. In Fig. 6.15(b) the product BH is plotted against B for the ferromagnetic alloy whose demagnetization curve is given in Fig. 6.15(a). It is evident that the optimum BH value occurs at point P and that the best material for a high-field permanent magnet is one with both a large remanence B_r and large coercivity H_c. To preserve the magnetization it is usual to place a 'keeper' of soft iron in the gap when the magnet is not in use. Since soft iron has a high permeability, (6.31) shows that it reduces the demagnetizing field H_m almost to zero.

The magnetic energy density in the gap is, by (6.14)

$$u = \frac{1}{2}\frac{B_g^2}{\mu_0}$$

By the principle of virtual work, the attractive force per unit area between the pole faces of a permanent magnet is just

$$\frac{F}{S_g} = \frac{1}{2}\frac{B_g^2}{\mu_0}$$

The construction of a permanent magnet must therefore be rigid enough to keep the air gap constant and resist this magnetic stress.

7

Maxwell's equations

Electromagnetic theory is a triumph of classical physics. It was completed in a set of differential equations by Maxwell between 1855 and 1865. In this chapter we collect together the basic equations of electromagnetism and express them in the form first devised by Maxwell to represent correctly the relationships between the electric field $\boldsymbol{E}$ and the magnetic field $\boldsymbol{B}$ in the presence of electric charges and electric currents, whether steady or rapidly fluctuating, in a vacuum or in matter. These are *Maxwell's equations.*

We shall find that we have already discussed three of these equations, due to Gauss and Faraday, but the Ampère's law has to be modified to allow for varying electric fields, whence it becomes known as Maxwell's law. We conclude this chapter with comments on the differences between statics and dynamics in electromagnetism and on the form that Maxwell's equations take in free space and in matter.

7.1 GAUSS'S AND FARADAY'S LAWS

Maxwell's equations express the fluxes and circulations of the electric and magnetic field vectors in differential form. The first one is derived from Gauss's law (2.9) for the electric field,

$$\int_S \boldsymbol{E}\cdot\mathbf{d}\boldsymbol{S} = \frac{1}{\varepsilon_0}\int_V \rho\,\mathrm{d}\tau$$

which with Gauss's divergence theorem (3.3) become

$$\boxed{\operatorname{div}\boldsymbol{E} = \rho/\varepsilon_0} \tag{7.1}$$

The second equation (4.28), also due to Gauss, was for the flux of the magnetic field,

$$\int_S \boldsymbol{B}\cdot\mathbf{d}\boldsymbol{S}=0$$

which in differential form (4.29) is

$$\boxed{\operatorname{div}\boldsymbol{B}=0} \tag{7.2}$$

The third equation (5.8), derived from Faraday's law of electromagnetic induction, is

$$\oint_C \boldsymbol{E}\cdot\mathbf{d}\boldsymbol{s}=-\int_S \frac{\partial \boldsymbol{B}}{\partial t}\cdot\mathbf{d}\boldsymbol{S}$$

which with Stokes's theorem (3.6) led to (5.9)

$$\boxed{\operatorname{curl}\boldsymbol{E}=-\frac{\partial \boldsymbol{B}}{\partial t}} \tag{7.3}$$

These equations are true for all electromagnetic fields, for moving charges as well as stationary charges, for high frequencies as well as steady states, although it is necessary to consider relativistic effects to prove this statement.

7.2 AMPÈRE'S AND MAXWELL'S LAWS

It would be logical at this point to expect that the fourth Maxwell equation would be derived from Ampère's law (4.24)

$$\oint_C \boldsymbol{B}\cdot\mathbf{d}\boldsymbol{s}=\mu_0\int_S \boldsymbol{j}\cdot\mathbf{d}\boldsymbol{S}$$

which correctly represents the circulation of the magnetic field for steady currents as μ_0 times the current density flux enclosed. But Maxwell noticed that when he applied this equation to the charging of a capacitor it did not work: the circulation around a contour C depended on where the imaginary surface S was drawn (Fig. 7.1). For the plane surface S_1, which cuts the current I, (4.24) gives

$$\oint_C \boldsymbol{B}\cdot\mathbf{d}\boldsymbol{s}=\mu_0 I$$

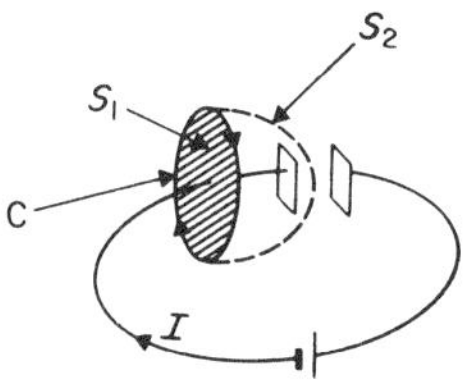

Fig. 7.1 A displacement current must exist in the conductor.

but a second surface S_2 for the same contour C does not cut the current and so for this surface

$$\oint_C \boldsymbol{B}\cdot\mathbf{d}\boldsymbol{s} = 0!$$

Maxwell argued that there must be a *displacement current* in the space between the capacitor plates which was equivalent to the external current. In a parallel plate capacitor the electric field $E = \sigma/\varepsilon_0$ (2.18) and so if the plates are of area A,

$$E = \frac{Q}{\varepsilon_0 A}$$

where Q is the charge on the plates. As the capacitor is charged the charging current is

$$I = \frac{\partial Q}{\partial t} = \varepsilon_0 A \frac{\partial E}{\partial t}$$

Maxwell therefore modified Ampère's law by adding the flux of this displacement current giving:

$$\oint_C \boldsymbol{B}\cdot\mathbf{d}\boldsymbol{s} = \mu_0\left(I + \varepsilon_0 A \frac{\partial E}{\partial t}\right)$$

Then for surface S_1 the circulation is $\mu_0 I$ and for surface S_2 it is

$$\mu_0\varepsilon_0 A \frac{\partial E}{\partial t} = \mu_0 I$$

and so constant. In general, therefore,

$$\oint_C \boldsymbol{B}\cdot\mathbf{d}\boldsymbol{s} = \mu_0\left\{\int_S \boldsymbol{j}\cdot\mathbf{d}\boldsymbol{S} + \int_S \varepsilon_0 \frac{\partial \boldsymbol{E}}{\partial t}\cdot\mathbf{d}\boldsymbol{S}\right\}$$

is the integral form of Maxwell's law, which, with Stokes's theorem, becomes the fourth Maxwell equation:

$$\boxed{\operatorname{curl} \boldsymbol{B} = \mu_0\left(\boldsymbol{j} + \varepsilon_0 \frac{\partial \boldsymbol{E}}{\partial t}\right)} \tag{7.4}$$

All electromagnetic fields satisfy these four Maxwell equations and with the electromagnetic force law (4.9),

$$\boxed{\boldsymbol{F} = q\boldsymbol{E} + q\boldsymbol{v} \times \boldsymbol{B}} \tag{7.5}$$

they summarize the whole of electromagnetism.

The electric and magnetic fields in Maxwell's equations refer to a classical 'point', which is conceived as an infinitesimal volume of a macroscopic field, but containing a very large number of atoms. In matter therefore the fields $\boldsymbol{E}$ and $\boldsymbol{B}$, and the densities ρ and $\boldsymbol{j}$, are averages over large numbers of microscopic particles (electrons, protons, neutrons). The equations are not limited to linear, isotropic media, but apply also to nonlinear, anistropic and non-homogeneous media.

7.3 STATICS AND DYNAMICS

We began electromagnetism by stating Coulomb's law and developing electrostatics. How many of our earlier equations are true for electrodynamics? Well, Coulomb's law obviously is not, for as soon as the charges move there is an electromagnetic force on them and we must use (7.5). Then the concept of the electric field as the gradient of a scalar potential (2.17)

$$\boldsymbol{E} = -\operatorname{grad} \phi$$

is only true for static fields, since it is based on the circulation law (2.10)

$$\oint_C \boldsymbol{E} \cdot \mathbf{d}\boldsymbol{s} = 0$$

which we saw in Chapter 5 became Faraday's law of electromagnetic induction (5.8)

$$\oint_C \boldsymbol{E} \cdot \mathbf{d}\boldsymbol{s} = -\int_S \frac{\partial \boldsymbol{B}}{\partial t} \cdot \mathbf{d}\boldsymbol{S}$$

the integral form of Maxwell's third equation.

It follows that Laplace's and Poisson's equations for $\nabla^2\phi$ are only valid for electrostatic fields and have to be modified in electrodynamics. On the other hand, the expressions that we have derived in electrostatics and in magnetism for electric and magnetic energy are true functions of the electromagnetic field at all frequencies. That is, the energy of an electromagnetic field, given by (2.36), is

$$U=\frac{1}{2}\int_\tau \boldsymbol{D}\cdot\boldsymbol{E}\,\mathrm{d}\tau$$

plus, for the magnetic energy (6.15),

$$U=\frac{1}{2}\int_\tau \boldsymbol{B}\cdot\boldsymbol{H}\,\mathrm{d}\tau$$

Of course such a statement has to be justified and this is given in the next chapter.

Finally, as Maxwell showed, the beautiful picture of a conductor as an equipotential surface with no electric field inside it is only true for electrostatics. In a conductor electrodynamic fields produce currents and so cause ohmic losses. The simple pictures of capacitors having only a capacitance C and of inductors having only an inductance L break down at high frequencies, even if the inductor is wound with superconducting wire and an excellent dielectric like mica is used in the capacitor.

7.4 FREE SPACE AND MATTER

In completely empty, or free space there can be no electric charges and no electric currents. So, Maxwell's equations (7.1) to (7.4) become:

$$\operatorname{div}\boldsymbol{E}=0 \tag{7.6}$$

$$\operatorname{div}\boldsymbol{B}=0 \tag{7.7}$$

$$\operatorname{curl}\boldsymbol{E}=-\frac{\partial\boldsymbol{B}}{\partial t} \tag{7.8}$$

$$\operatorname{curl}\boldsymbol{B}=\mu_0\varepsilon_0\frac{\partial\boldsymbol{E}}{\partial t} \tag{7.9}$$

The surprising result of these equations, as Maxwell first showed in 1864, is that electric and magnetic fields do not merely exist in free space, but can propagate at the speed of light over galactic distances. So using satellites modern astronomy is able to explore the universe over the entire electromagnetic spectrum from cosmic rays to long-wavelength radio waves (Appendix D). We shall solve (7.6) to (7.9) for the electric and magnetic fields of electromagnetic waves in Chapter 9.

In the presence of matter, many physicists prefer to reformulate Maxwell's equations (7.1) to (7.4) in terms of the four fields $\boldsymbol{E}$, $\boldsymbol{D}$, $\boldsymbol{B}$ and $\boldsymbol{H}$, where the electric displacement $\boldsymbol{D}$ and the magnetizing field $\boldsymbol{H}$, defined by (2.30) and (6.7), are as follows:

$$\boldsymbol{D} = \varepsilon_0 \boldsymbol{E} + \boldsymbol{P}$$

and

$$\boldsymbol{H} = \frac{\boldsymbol{B}}{\mu_0} - \boldsymbol{M}$$

Here $\boldsymbol{P}$ is the electric polarization in a dielectric medium and $\boldsymbol{M}$ is the magnetization in magnetic matter. The result is that (7.1) and (7.4) are changed, but (7.2) and (7.3), which do not contain any sources, remain as before. We will now show explicitly how first (7.1) and then (7.4) can be rewritten in terms of $\boldsymbol{D}$ and $\boldsymbol{H}$ for use in dielectrics and magnetic matter.

When a dielectric medium is present the charge density ρ in (7.1) is the sum of the density ρ_p of any polarization charges and the density ρ_f of any free charges. Therefore, (7.1) becomes

$$\operatorname{div} \varepsilon_0 \boldsymbol{E} = \rho_p + \rho_f \tag{7.10}$$

We saw in Chapter 2 that the flux of $\boldsymbol{P}$ is given by a type of Gauss's law for polarized dielectrics:

$$\int_S \boldsymbol{P}\cdot\mathbf{d}\boldsymbol{S} = -\int_V \rho_p \, d\tau \tag{7.11}$$

Applying Gauss's divergence theorem (Appendix E) to this equation we have

$$\int_V \operatorname{div} \boldsymbol{P} \, d\tau = -\int_V \rho_p \, d\tau$$

and so

$$\operatorname{div} \boldsymbol{P} = -\rho_p \tag{7.12}$$

Substituting for ρ_p in (7.10) gives

$$\operatorname{div}(\varepsilon_0 \boldsymbol{E} + \boldsymbol{P}) = \rho_f$$

or

$$\operatorname{div} \boldsymbol{D} = \rho_f \tag{7.13}$$

The fourth Maxwell equation, (7.4), includes a term $\partial \boldsymbol{E}/\partial t$ for electric fields that are varying with time. In the presence of such time-dependent fields the motion of the polarization charges in a dielectric produces a polarization current of density $\boldsymbol{j}_p$. Since charge is conserved, the outward flux of such a current density from a volume V must be equal to the rate of decrease of the polarization charges within it:

$$\int_S \boldsymbol{j}_p \cdot \mathbf{d}\boldsymbol{S} = -\frac{\partial}{\partial t}\int_V \rho_p \, d\tau \tag{7.14}$$

From (7.11) this becomes

$$\int_S \boldsymbol{j}_p \cdot \mathbf{d}\boldsymbol{S} = \frac{\partial}{\partial t}\int_S \boldsymbol{P} \cdot \mathbf{d}\boldsymbol{S}$$

and, since the time derivative can be taken either before or after the integration,

$$\boldsymbol{j}_p = \frac{\partial \boldsymbol{P}}{\partial t} \tag{7.15}$$

Applying (7.4) to a polarizable and magnetizable medium we must put

$$\boldsymbol{j} = \boldsymbol{j}_f + \boldsymbol{j}_p + \boldsymbol{j}_m \tag{7.16}$$

where the total electric current density $\boldsymbol{j}$ is the sum of the conduction current density $\boldsymbol{j}_f$ due to the free charges ρ_f, the polarization current density $\boldsymbol{j}_p$ due to the polarization charges ρ_p, and the magnetization current density $\boldsymbol{j}_m$ associated with magnetized matter. This arises from the atomic currents inside the matter which are equivalent to small magnetic dipoles. As we have seen in Chapter 6, the *uniform* magnetization $\boldsymbol{M}$ of a block can thus be replaced by an equivalent magnetization surface current density i_m which acts in the direction given by

$$\boldsymbol{M} \times \hat{\mathbf{n}} = i_m \tag{7.17}$$

where $\hat{\mathbf{n}}$ is the outward normal of the surface of the block containing the current. In this case the volume current density $\boldsymbol{j}_m$ is zero and there is only a surface current density $\boldsymbol{i}_m$.

For a nonuniformly magnetized material, however, there is also an equivalent current density $\boldsymbol{j}_m$, by analogy with Ampère's law:

$$\oint_C \frac{\boldsymbol{B}}{\mu_0} \cdot \mathbf{d}\boldsymbol{s} = \int_S \boldsymbol{j} \cdot \mathbf{d}\boldsymbol{S} \tag{7.18}$$

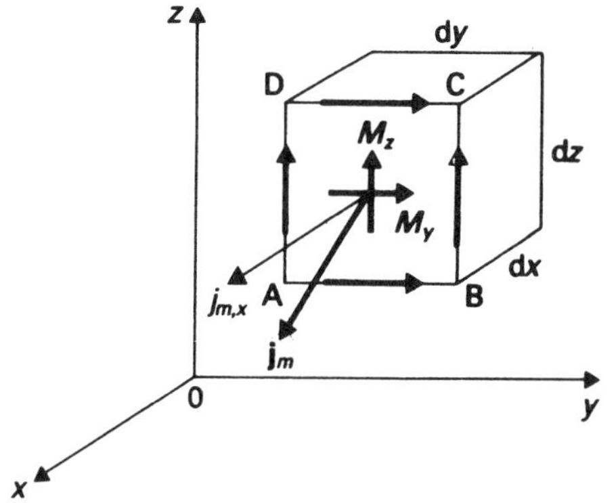

Fig. 7.2 An elementary volume of nonuniform magnetization in equivalent to a volume of current density $\boldsymbol{j}_\mathrm{m}$.

namely

$$\oint_C \boldsymbol{M}\cdot\mathbf{d}\boldsymbol{s} = \int_S \boldsymbol{j}_\mathrm{m}\cdot\mathbf{d}\boldsymbol{S} \tag{7.19}$$

This is illustrated in Fig. 7.2 where, for the contour ABCD,

$$\begin{aligned}\oint_C \boldsymbol{M}\cdot\mathbf{d}\boldsymbol{s} &= \left(M_y - \frac{\partial M_y}{\partial z}\frac{\mathrm{d}z}{2}\right)\mathrm{d}y + \left(M_z + \frac{\partial M_z}{\partial y}\frac{\mathrm{d}y}{2}\right)\mathrm{d}z \\ &\quad - \left(M_y + \frac{\partial M_y}{\partial z}\frac{\mathrm{d}z}{2}\right)\mathrm{d}y - \left(M_z - \frac{\partial M_z}{\partial y}\frac{\mathrm{d}y}{2}\right)\mathrm{d}z \\ &= -\frac{\partial M_y}{\partial z}\mathrm{d}y\,\mathrm{d}z + \frac{\partial M_z}{\partial y}\mathrm{d}y\,\mathrm{d}z \\ &= (\mathrm{curl}\,\boldsymbol{M})_x \mathrm{d}y\,\mathrm{d}z.\end{aligned}$$

However for this contour the flux of $\boldsymbol{j}_\mathrm{m} = j_{\mathrm{m}x}\,\mathrm{d}y\,\mathrm{d}z$, so that

$$j_{\mathrm{m},x} = (\mathrm{curl}\,\boldsymbol{M})_x$$

Similarly for the other faces of the cube $j_{\mathrm{m},y} = (\mathrm{curl}\,\boldsymbol{M})_y$ and $j_{\mathrm{m},z} = (\mathrm{curl}\,\boldsymbol{M})_z$, and the equivalent current density is

$$j_\mathrm{m} = \mathrm{curl}\,\boldsymbol{M} \tag{7.20}$$

Combining (7.15), (7.16) and (7.20), we have for the total current density $\boldsymbol{j}$ of (7.4)

$$j = j_f + \frac{\partial \boldsymbol{P}}{\partial t} + \text{curl}\,\boldsymbol{M} \tag{7.21}$$

and so Maxwell's fourth equation becomes

$$\text{curl}\left(\frac{\boldsymbol{B}}{\mu_0}\right) = j_f + \frac{\partial \boldsymbol{P}}{\partial t} + \text{curl}\,\boldsymbol{M} + \varepsilon_0 \frac{\partial \boldsymbol{E}}{\partial t}$$

or

$$\text{curl}\left(\frac{\boldsymbol{B}}{\mu_0} - M\right) = j_f + \left(\varepsilon_0 \frac{\partial \boldsymbol{E}}{\partial t} + \frac{\partial \boldsymbol{P}}{\partial t}\right)$$

From the definitions of $\boldsymbol{D}$ and $\boldsymbol{H}$ in (2.30) and (6.7) it therefore follows that

$$\text{curl}\,\boldsymbol{H} = j_f + \frac{\partial \boldsymbol{D}}{\partial t} \tag{7.22}$$

We conclude that for a polarizable, magnetizable medium the four Maxwell equations can also be written as follows:

$$\text{div}\,\boldsymbol{D} = \rho_f \tag{7.13}$$

$$\text{div}\,\boldsymbol{B} = 0 \tag{7.2}$$

$$\text{curl}\,\boldsymbol{E} = -\frac{\partial \boldsymbol{B}}{\partial t} \tag{7.3}$$

$$\text{curl}\,\boldsymbol{H} = j_f + \frac{\partial \boldsymbol{D}}{\partial t} \tag{7.22}$$

and we shall use either this set or (7.1) to (7.4) as appropriate. Both sets of equations apply to nonlinear, anistropic, nonhomogeneous media, providing the definitions for $\boldsymbol{D}$ and $\boldsymbol{H}$ are those given in (2.30) and (6.7). The alternative definitions, in terms of the dimensionless relative permittivity ε_r and relative permeability μ_r, are

$$\boldsymbol{B} = \mu_r \mu_0 \boldsymbol{H} \tag{7.23}$$

$$\boldsymbol{D} = \varepsilon_r \varepsilon_0 \boldsymbol{E} \tag{7.24}$$

are particularly useful in linear, isotropic materials. Then the magnetization $\boldsymbol{M}$ is proportional to $\boldsymbol{H}$ and the polarization $\boldsymbol{P}$ is proportional to $\boldsymbol{E}$ and so μ_r, ε_r are scalars. In anisotropic materials, such as piezoelectric crystals, the permittivity becomes a tensor,

$$\varepsilon_{\alpha\beta} = \frac{1}{\varepsilon_0} \frac{\partial D_\alpha}{\partial E_\beta} \tag{7.25}$$

while in nonlinear materials, such as ferromagnetics, the permeability varies over the hysteresis loop and becomes a differential,

$$\mu_r(B, H) = \frac{1}{\mu_0}\frac{dB}{dH} \tag{7.26}$$

An important principle in electromagnetism is the conservation of electric charge. It is recognized as of similar validity to the principles of conservation of mass–energy and of momentum throughout physics. In integral form, in (4.2), we had

$$\int_S \boldsymbol{j}\cdot\mathbf{d}\boldsymbol{S} = \int \frac{dq}{dt}$$

where the charge q is flowing out of a surface of area S. If S is a closed surface and the charge density inside is ρ, then this can be written

$$\int_S \boldsymbol{j}\cdot\mathbf{d}\boldsymbol{S} = -\frac{d}{dt}\left(\int_V \rho\, d\tau\right) \tag{7.27}$$

By applying Gauss's divergence theorem (Appendix E) this becomes

$$\int_S \boldsymbol{j}\cdot\mathbf{d}\boldsymbol{S} = \int_V \operatorname{div}\boldsymbol{j}\, d\tau \tag{7.28}$$

If we apply (7.27) and (7.28) to a small volume $d\tau$, then we obtain the *equation of continuity*:

$$\boxed{\operatorname{div}\boldsymbol{j} = -\frac{\partial\rho}{\partial t}} \tag{7.29}$$

This equation is not, however, additional to Maxwell's equations, because it is inherent in them.

The simplest way of showing that Maxwell's equations contain the principle of conservation of charge is by combining (7.13) and (7.22). From (7.22)

$$\operatorname{div}\operatorname{curl}\boldsymbol{H} = \operatorname{div}\boldsymbol{j}_f + \operatorname{div}\frac{\partial\boldsymbol{D}}{\partial t} = 0$$

since the divergence of a curl of any vector is identically zero. Therefore

$$\operatorname{div}\boldsymbol{j}_f = -\operatorname{div}\left(\frac{\partial\boldsymbol{D}}{\partial t}\right) = -\frac{\partial\rho_f}{\partial t}$$

from (7.13) and we recover (7.29) for the free charge density ρ_f and the conduction current density $\boldsymbol{j}_f$. In a similar way it can be shown that (7.4), where $\boldsymbol{j}$ is the current density in matter and can include a polarization current density $\partial \boldsymbol{P}/\partial t$ and a magnetization current density curl $\boldsymbol{M}$, leads to (7.29) for the total charge density

$$\rho = \rho_f + \rho_p$$

To solve Maxwell's equations for $\boldsymbol{E}$ and $\boldsymbol{B}$ we have to integrate them for the proper boundary conditions. This is simplest in free space, but solutions can be found for dielectrics, semiconductors, conductors and indeed for any crystalline or amorphous material. The solutions include the simplest examples with high symmetry that we have discussed for distributions of static charges and steady currents. However, it was Maxwell's deduction that electromagnetic waves could be generated by high frequency currents and the discovery of such radiation by Hertz that first convinced everyone that Maxwell's equations did contain the whole of electromagnetism. They form the starting point of the remaining chapters of this book.

8

Electromagnetism and relativity

In this chapter we show first that the magnetic field $\boldsymbol{B}$ arises naturally from a Lorentz transformation of the electric force in Coulomb's law and so we can investigate the electric and magnetic fields of rapidly moving charges. Then by introducing a vector potential $\boldsymbol{A}$ subject to the Lorentz gauge condition we find a remarkable similarity between the equations of electrostatics, expressed in terms of the scalars ϕ (potential) and ρ, and the equations of magnetostatics in terms of the vectors $\boldsymbol{A}$ and $\boldsymbol{j}$. Finally we show, using a four-dimensional notation, that it is possible to reduce Maxwell's four equations to just one equation which expresses directly the invariance of electromagnetism under the Lorentz transformation.

8.1 LORENTZ TRANSFORMATIONS

In Einstein's special theory of relativity Newton's laws of motion are modified so that instead of a Galilean transformation, valid for mechanical phenomena when relative speeds are much less than the speed of light, c, a Lorentz transformation valid for all types of physical phenomena at all speeds is required. As a consequence of this all physical laws must be *invariant* under a Lorentz transformation.

A simple example to show the difference between Galilean and Lorentz transformations is to consider the coordinates of a point P′ in an inertial frame S' (x', y', z') moving with speed u (parallel to $0x$) relative to an inertial frame S (x, y, z), as shown in Fig. 8.1(a). The transformations are then as follows:

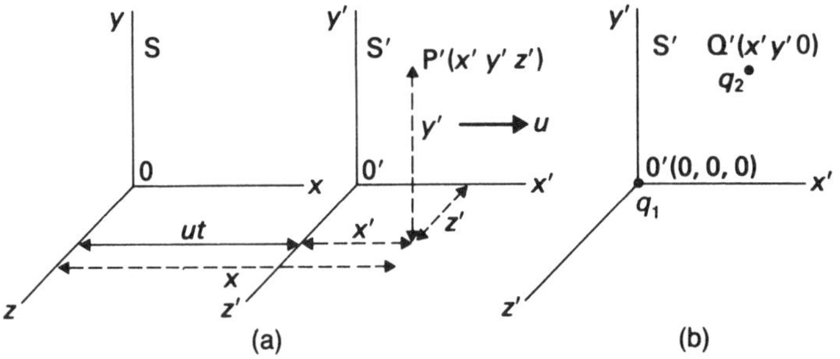

Fig. 8.1 Frame S' (x', y', z') is moving at speed u along Ox and has moved a distance ut from the S frame (x, y, z) during the time t. The coordinates of P′ are (x', y', z') and Q′ $(x', y', 0)$.

Galilean transformations	*Lorentz transformations*	
$x' = x - ut$	$x' = \gamma(x - ut)$	
$y' = y$	$y' = y$	(8.1)
$z' = z'$	$z' = z$	
$t' = t$	$t' = \gamma\{t - (u/c^2)x\}$	

where $\gamma = 1/\{1 - (u^2/c^2)\}^{1/2} = 1/(1 - \beta^2)^{1/2} \geqslant 1$. As a result of applying relativity theory to an object moving with a speed u relative to an observer the following transformations are found:

1. the FitzGerald contraction of length L_0 to $L = L_0/\gamma$;
2. the dilatation of a time interval T_0 to $T = \gamma T_0$;
3. the increase of a rest mass m_0 to $m = \gamma m_0$.

From these it follows that relativistic energy $U = \gamma m_0 c^2$, relativistic momentum $\boldsymbol{p} = (\gamma m_0)\, \boldsymbol{v}$ and that a force $\boldsymbol{F}$ in a frame S transforms to a force $\boldsymbol{F}' = \gamma \boldsymbol{F}$ in a moving frame S'.

The basic quantities in electromagnetism are electric charge q, electric charge density ρ and electric current density $\boldsymbol{j}$. What happens to these quantities when we observe them in a moving frame? We might expect q to transform to γq like mass, or to q/γ like a length. In fact this does not occur. Electric charge, unlike mass, is *invariant* at all speeds. The best evidence for this comes from measurements of the ratio of the electronic charge-to-mass ratio, e/m, for high-energy electrons from accelerators that operate at GeV energies, where the electron's speed can be within 1 part in 10^8 of c. When the increase of mass from m_0 to γm_0 is allowed for, there is no change in the value of e.

In order to find how charge density ρ and current density $\boldsymbol{j} = \rho\boldsymbol{v}$ transform, we require the transformations for an element of volume $d\tau_0$ and a velocity. For simplicity let us assume we place $d\tau_0$ at P′ in the S' frame (Fig. 8.1(a)) and allow it to move at a speed v'_x in this frame parallel to the x' axis. It can readily be shown from the Lorentz transformations of length and time in (8.1) that v'_x is related to v_x in the S frame by

$$v'_x = \frac{v_x - u}{1 - (v_x u/c^2)} \tag{8.2}$$

In the S' frame the volume element will contract only in the x direction so that it becomes

$$d\tau' = d\tau_0 \left\{1 - \left(\frac{v'_x}{c}\right)^2\right\}^{1/2}$$

while in the S frame it will appear to be

$$d\tau = d\tau_0 \left\{1 - \left(\frac{v_x}{c}\right)^2\right\}^{1/2}$$

Therefore

$$\frac{d\tau'}{d\tau} = \left\{\frac{1 - \left(\dfrac{v'_x}{c}\right)^2}{1 - \left(\dfrac{v_x}{c}\right)^2}\right\}^{1/2}$$

and when v'_x is substituted from (8.2) the ratio simplifies to

$$\frac{d\tau'}{d\tau} = \frac{1}{\gamma\{1 - (v_x u/c^2)\}} \tag{8.3}$$

Using (8.2) and (8.3) we can now transform a charge density ρ' in the S' frame to a charge density ρ in the S frame. The charge q will be invariant, so that

$$q = \rho\, d\tau = \rho'\, d\tau'$$

and therefore

$$\frac{\rho'}{\rho} = \frac{d\tau}{d\tau'} = \gamma\left(1 - \frac{v_x u}{c^2}\right)$$

This can more usefully be expressed in terms of the x-component of the current density $\boldsymbol{j}$, since $j_x = \rho v_x$ and so

$$\rho' = \gamma\left(\rho - \frac{u j_x}{c^2}\right) \tag{8.4}$$

On the other hand

$$j'_x = \rho' v'_x = \gamma\left(\rho - \frac{u j_x}{c^2}\right)\left\{\frac{v_x - u}{1 - (v_x u/c^2)}\right\}$$

and this simplifies to

$$j'_x = \gamma(j_x - u\rho) \tag{8.5}$$

Since the only component of $\boldsymbol{v}$ is v_x, then

$$j'_y = j_y \text{ and } j'_z = j_z \tag{8.6}$$

When we compare (8.4), (8.5) and (8.6) with the Lorentz transformations for x', y', z' and t' in (8.1), we see that the current density $\boldsymbol{j}$ and charge density ρ transform just like the position vector $\boldsymbol{r}$ and the time t. Another way of putting it is that the four-vector $j_\nu \equiv (\boldsymbol{j}, ic\rho)$ and its invariant $j_\nu^2 = j^2 - c^2\rho^2$ in electromagnetism correspond to the four-vector $\boldsymbol{r}_\nu = (\boldsymbol{r}, ict)$ and its invariant $r_\nu^2 = r^2 - c^2 t^2$ in mechanics, where we denote the Einstein summation convention over four coordinates by the Greek letter ν (for example, $j_\nu^2 \equiv j_1^2 + j_2^2 + j_3^2 + j_4^2$, where $j_1 = j_x, j_2 = j_y, j_3 = j_z, j_4 = ic\rho$).

8.2 FIELDS OF MOVING CHARGES

8.2.1 Origin of *B*

We are now in a position to see what happens when we transform the Coulomb force between two charges at rest in a moving frame S' back into the laboratory frame S. Let us put one charge q_1 at 0′ in Fig. 8.1(b) and the other q_2 at Q′ so that their coordinates in the S' frame are q_1 (0, 0, 0) and q_2 $(x', y', 0)$. In this frame the Coulomb force $\boldsymbol{F}'$ has components given by

$$F'_x = \frac{q_1 q_2 x'}{4\pi\varepsilon_0 r'^3}, \quad F'_y = \frac{q_1 q_2 y'}{4\pi\varepsilon_0 r'^3}, \quad F'_z = 0 \tag{8.7}$$

To simplify the calculation of the force $\boldsymbol{F}$ in the S frame we will compute it at time $t = 0$, when q_1 is at 0 and the two frames coincide.

Since the relative motion of S' with respect to S is along 0x, only the y and z components of F' are contracted and

$$F_x = F'_x, \quad F_y = F'_y/\gamma, \quad F_z = F'_z/\gamma \tag{8.8}$$

The force $\boldsymbol{F}$ therefore has components given by

$$F_x = \frac{q_1 q_2 x'}{4\pi\varepsilon_0(x'^2 + y'^2)^{3/2}} \quad \text{and} \quad F_y = \frac{q_1 q_2 y'}{4\pi\varepsilon_0\gamma(x'^2 + y'^2)^{3/2}}$$

Substituting for x', y' at $t = 0$ from (8.1) and multiplying by γ, these become

$$F_x = \frac{\gamma q_1 q_2 x}{4\pi\varepsilon_0(\gamma^2 x^2 + y^2)^{3/2}}, \quad F_y = \frac{\gamma q_1 q_2 y}{4\pi\varepsilon_0(\gamma^2 x^2 + y^2)^{3/2}}\left(1 - \frac{u^2}{c^2}\right)$$

At $t = 0$, q_2 is at $\boldsymbol{r} = x\hat{\mathbf{i}} + y\hat{\mathbf{j}}$, so we can put (writing $x\hat{\mathbf{i}} = \boldsymbol{r} - y\hat{\mathbf{j}}$)

$$\boldsymbol{F} = \frac{\gamma q_1 q_2 \boldsymbol{r}}{4\pi\varepsilon_0(\gamma^2 x^2 + y^2)^{3/2}} - \frac{\gamma q_1 q_2 u^2 y\hat{\mathbf{j}}}{4\pi\varepsilon_0 c^2(\gamma^2 x^2 + y^2)^{3/2}}$$

Alternatively, since $\boldsymbol{u} = u\hat{\mathbf{i}}$ and $\hat{\mathbf{i}} \times \hat{\mathbf{k}} = -\hat{\mathbf{j}}$, we can write

$$\boldsymbol{F} = q_2\left\{\frac{\gamma q_1 \boldsymbol{r}}{4\pi\varepsilon_0(\gamma^2 x^2 + y^2)^{3/2}} + \frac{\boldsymbol{u} \times \gamma q_1 u y\hat{\mathbf{k}}}{4\pi\varepsilon_0 c^2(\gamma^2 x^2 + y^2)^{3/2}}\right\} \tag{8.9}$$

This equation is just the Lorentz force law

$$\boldsymbol{F} = q_2(\boldsymbol{E} + \boldsymbol{u} \times \boldsymbol{B}) \tag{8.10}$$

where

$$\boldsymbol{E} = \frac{\gamma q_1 \boldsymbol{r}}{4\pi\varepsilon_0(\gamma^2 x^2 + y^2)^{3/2}} \quad \text{and} \quad \boldsymbol{B} = \frac{\mu_0 \gamma q_1 u y\hat{\mathbf{k}}}{4\pi(\gamma^2 x^2 + y^2)^{3/2}} \tag{8.11}$$

These expressions are readily seen to have the familiar form,

$$E_0 = \left(\frac{q}{4\pi\varepsilon_0 r^2}\right)\hat{\mathbf{r}} \quad \text{and} \quad \boldsymbol{B}_0 = \left(\frac{\mu_0}{4\pi}\right)\left(\frac{\hat{\mathbf{j}} \times \hat{\mathbf{r}}}{r^2}\right)$$

when u is small and $\gamma \cong 1$. The relationships between $\boldsymbol{F}$, $\boldsymbol{E}$ and $\boldsymbol{B}$ are shown in Fig. 8.2.

This straightforward application of the special theory of relativity has therefore shown that an electric force $\boldsymbol{F}'$ in the moving frame is transformed into the Lorentz force $\boldsymbol{F}$ in the laboratory frame. The magnetic field $\boldsymbol{B}$, which is associated with moving charges and electric currents in elementary theory, is not a separate phenomenon but a relativistic transformation of a moving electric field. We will now

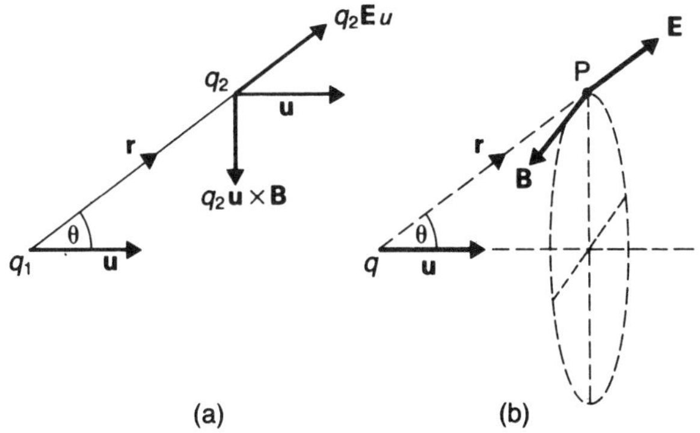

Fig. 8.2 The force $\boldsymbol{F}$ between two charges q_1 and q_2 moving at the same velocity $\boldsymbol{u}$ is the vector sum of an electric force $q_2\boldsymbol{E}_u$ and a magnetic force $q_2\boldsymbol{u} \times \boldsymbol{B}$. (b) The electric field $\boldsymbol{E}$ and magnetic field $\boldsymbol{B}$ at $P(r, \theta)$ due to a charge q moving at velocity $\boldsymbol{u}$.

consider how these fields change when charges are travelling at relativistic speeds.

8.2.2 Electric fields

The electric field $\boldsymbol{E}_u$ of a charge moving at speed u is radial (Fig. 8.2(a)), just as for a stationary charge, but its magnitude is a function of θ that depends on the ratio $\beta = u/c$. This is readily seen for the limiting cases of $\theta = 0$ and $\pi/2$. From (8.11) and Fig. 8.2(a), when $\theta = 0$, $x = r$, $y = 0$ and so the magnetic field is zero and the electric field is

$$\boldsymbol{E}_u = \frac{1}{\gamma^2}\boldsymbol{E}_0$$

Clearly as the speed u increases towards c, so the electric field also tends to zero. On the other hand for $\theta = \pi/2$, $x = 0$, $y = r$ and the electric field is

$$\boldsymbol{E}_u = \gamma \boldsymbol{E}_0$$

and this increases as u increases towards c. At this angle the electric and magnetic forces are in direct opposition and as u increases towards c the total force (Fig. 8.2(a)) on q_2 due to q_1 tends to zero. The changes in E_u (θ) at other angles as u increases are shown in Fig. 8.3(a) for several values of $\beta = u/c$. It is seen that the effect is

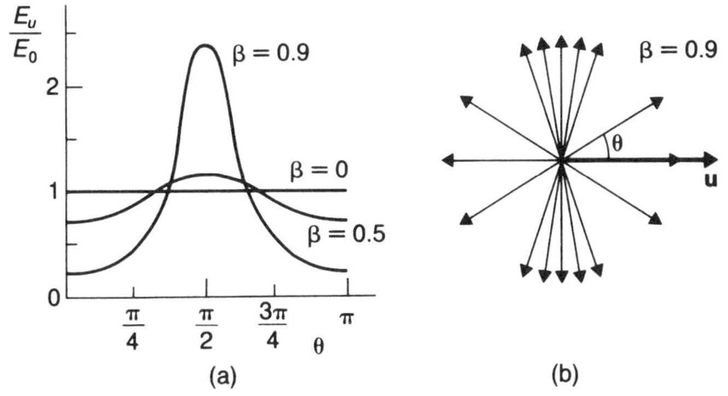

Fig. 8.3 (a) The electric field E_u/E_0 at angle θ to velocity $\boldsymbol{u}$ of charge, as seen by a stationary observer, for $\beta = u/c = 0$, 0.5, and 0.9. (b) Lines of the electric field $\boldsymbol{E}_u$ from a charge at velocity $\boldsymbol{u}$ and $\beta = 0.9$, as seen by a stationary observer.

significant when u is $0.5c$, the speed an electron reaches when accelerated through 80 keV. At the higher speed of $0.9c$ a stationary observer would find a distinctly non-uniform, radial electric field, as shown in Fig. 8.3(b). Ultimately this field exists only in a thin disc normal to the direction of motion of the charge.

8.2.3 Magnetic fields

The magnetic field $\boldsymbol{B}$ of a moving charge is always normal to the velocity $\boldsymbol{u}$ of the charge and so lines of $\boldsymbol{B}$ are circles centred on the trajectory of $\boldsymbol{u}$ (Fig. 8.4(a)). A stationary charge has no magnetic field and, as $\beta = u/c$ increases, the magnetic field at first increases at all angles (Fig. 8.4(b)) but for $\beta > 0.5$, the magnetic field becomes increasingly concentrated into a thin disc normal to the direction of motion of the charge. Thus at these high relativistic speeds both the electric and the magnetic fields are concentrated into the same plane normal to $\boldsymbol{u}$.

8.2.4 General case

So far we have only considered the case where the charges are both moving at the same speed, that of the moving frame. In general the velocity $\boldsymbol{u}$ of the moving frame (O′ relative to O in Fig. 8.5) will differ from the velocity $\boldsymbol{v}$ of a charge at P in the stationary (laboratory)

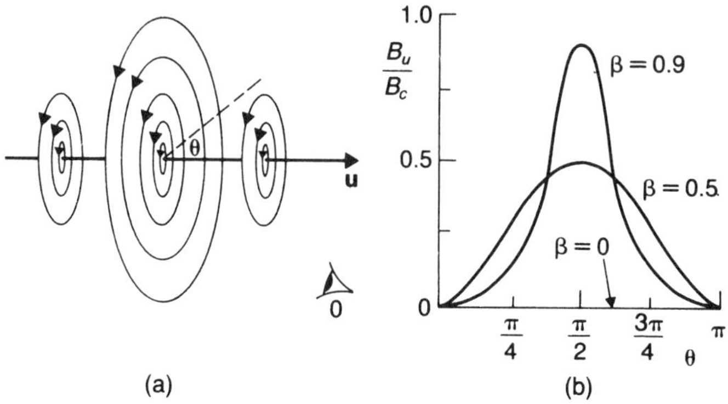

Fig. 8.4 (a) Lines of the magnetic field $\boldsymbol{B}_u$ from a charge at velocity $\boldsymbol{u}$ and speed $\beta = 0.5$, as seen by a stationary observer O. (b) The magnetic field $\boldsymbol{B}_u/\boldsymbol{B}_c$ at angle θ to velocity $\boldsymbol{u}$ of the charge, as seen by a stationary observer, for $\beta = 0$, 0.5 and 0.9 and $\boldsymbol{B}_c = \mu_0 qc/4\pi r^2$.

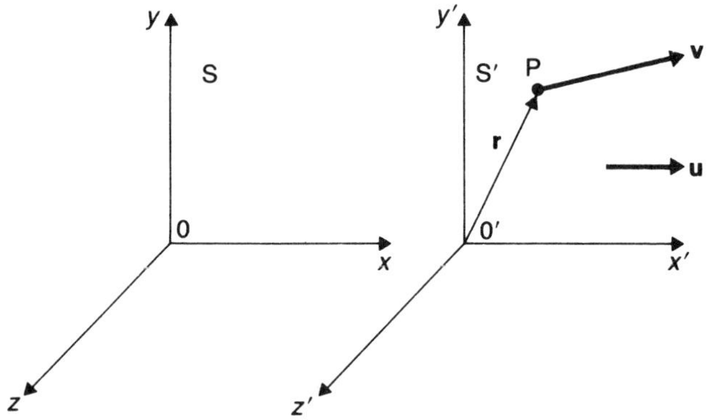

Fig. 8.5 The frame $\boldsymbol{S}'$ is moving at speed u along Ox relative to the laboratory frame $\boldsymbol{S}$ and its origin has moved from O to O′ in time t. The point P(x, y, z) is moving at a velocity $\boldsymbol{v}$ in the laboratory frame, where $\boldsymbol{v}$ differs in magnitude and direction from $\boldsymbol{u}$.

frame. In this case the distance O′P in the laboratory frame after time t is

$$\boldsymbol{r} = (x - ut)\hat{\mathbf{i}} + y\hat{\mathbf{j}} + z\hat{\mathbf{k}} \tag{8.12}$$

and, using the Lorentz transformations of (8.1), this distance in the moving frame is

$$\boldsymbol{r}' = \gamma(x - ut)\hat{\mathbf{i}} + y\hat{\mathbf{j}} + z\hat{\mathbf{k}} \tag{8.13}$$

In order to find the force law for this moving charge we consider, as before, the force exerted by a charge, q_1 at O′ (0, 0, 0) on a charge q_2 at P (x', y', z'), where the coordinates are in the moving frame where q_1 is stationary. In this frame the Coulomb force $\boldsymbol{F}'$ is then

$$\boldsymbol{F}' = \frac{q_1 q_2}{4\pi\varepsilon_0 r'^3}\{x'\hat{\mathbf{i}} + y'\hat{\mathbf{j}} + z'\hat{\mathbf{k}}\} \tag{8.14}$$

It is left as an exercise for the reader (exercise 8.1, Appendix G) using the Lorentz transformations for the coordinates, the velocities and the components of the force (Appendix F) to show that the Coulomb force $\boldsymbol{F}'$ transforms into the Lorentz force $\boldsymbol{F}$ in the laboratory frame:

$$\boldsymbol{F} = q_2\{\boldsymbol{E} + (\boldsymbol{v} \times \boldsymbol{B})\} \tag{8.15}$$

where

$$\boldsymbol{E} = \frac{\gamma q_1 \boldsymbol{r}}{4\pi\varepsilon_0 r'^3} \quad \text{and} \quad \boldsymbol{B} = \left(\frac{\mu_0}{4\pi}\right)\frac{\gamma q_1 u(-z\hat{\mathbf{j}} + y\hat{\mathbf{k}})}{r'^3} \tag{8.16}$$

By comparing (8.15) and (8.10) we see that the magnetic force in the laboratory frame due to q_1 on q_2 at P is proportional to the velocity $\boldsymbol{v}$ *in that frame*, as well as depending on the speed of q_1 through the factor γu. We conclude that the magnetic force is:

1. zero in a given frame of reference, unless $\boldsymbol{v}$ is finite in that frame;
2. independent of the component of $\boldsymbol{v}$ normal to $\boldsymbol{u}$;
3. normal to $\boldsymbol{v}$ and so does no work.

The last point is used extensively in particle physics when a magnet deflects a beam of charged particles without changing their kinetic energy.

8.2.5 Current in a wire

We can use the concepts of special relativity to see how a current I in a long, thin metal wire (Fig. 8.6) produces a force on a charge q moving with a velocity $\boldsymbol{v}$ relative to the wire. The wire is stationary in the laboratory (S) frame and in this frame there is charge neutrality, so that the linear charge density λ_p of the positive ions is exactly equal and opposite to the linear charge density λ_n of the conduction electrons, i.e.

$$\lambda_p + \lambda_n = 0 \tag{8.17}$$

The conduction electrons have a small drift velocity $\boldsymbol{u}$ along the wire

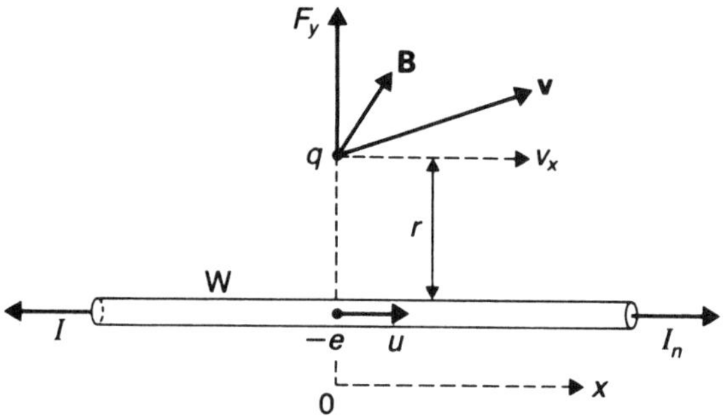

Fig. 8.6 Wire W carrying a conventional current I is at a distance r from a positive charge q moving with velocity $\boldsymbol{v}$. The conduction electrons drifting at speed $\boldsymbol{u}$ parallel to the x-axis produce an electron current I_n which exerts a force $F_y = qv_xB$ on q.

and are stationary in the moving (S') frame, which coincides with the S frame at $t = t' = 0$.

In the laboratory frame the positive ions exert a force F_y on q given by

$$F_y = qE_p = q\lambda_p/(2\pi\varepsilon_0 r) \tag{8.18}$$

where r is the radial distance of q from the wire. Similarly, in the moving frame the conduction electrons with linear density λ'_n exert a force F'_y given by

$$F'_y = q\lambda'_n/(2\pi\varepsilon_0 r) \tag{8.19}$$

since q, r are invariant. An observer in this frame would measure a total electronic charge of $\lambda'_n L$, where the length L has contracted from L_0 to L_0/γ by the FitzGerald contraction. Therefore,

$$\lambda'_n = \gamma\lambda_n \tag{8.20}$$

and

$$F'_y = q\gamma\lambda_n/(2\pi\varepsilon_0 r) \tag{8.21}$$

Using the Lorentz transformation for a force (Appendix F) this becomes

$$F_y = \frac{q\lambda_n}{2\pi\varepsilon_0 r}\left\{1 - \frac{uv_x}{c^2}\right\}$$

or

$$F_y = \frac{q}{2\pi\varepsilon_0 r}\left\{\lambda_n - \left(\frac{v_x}{c^2}\right)I_n\right\} \tag{8.22}$$

where $I_n = \lambda_n u$ is the electron current. Therefore, combining (8.18) and (8.22), the total force on q is

$$F_y = \frac{q}{2\pi\varepsilon_0 r}\left\{\lambda_p + \lambda_n - \frac{v_x I_n}{c^2}\right\}$$

and using (8.17) this is just

$$F_y = -\frac{qv_x I_n}{2\pi\varepsilon_0 c^2 r} = -\frac{\mu_0 q v_x I_n}{2\pi r} \tag{8.23}$$

The electric fields due to the positive ions and the conduction electrons cancel perfectly and the remaining force can be recognized as the Lorentz force $q\boldsymbol{v} \times \boldsymbol{B}$, where $\boldsymbol{B} = (\mu_0 I/2\pi r)\hat{\phi}$ for a conventional current $I = -\lambda_n u = -I_n$. We now see that the magnetic field of a current in a wire results from the relativistic transformation of the electric field of the moving electrons, despite the fact that typically the drift velocity of such electrons is about $10^{-4}\,\mathrm{m\,s^{-1}}$, so that $\gamma = 1$ to an accuracy of 1 part in 10^{25}! For example a current of 1 A in a copper wire of cross-sectional area $1\,\mathrm{mm}^2$ has a linear charge density of about

$$\lambda_n = -10^{29} \times 10^{-6} \times 1.6 \times 10^{-19} = -1.6 \times 10^4\,\mathrm{C\,m^{-1}}$$

while if q is an electron in a second wire and drift velocity $v_x = -10^{-4}\,\mathrm{m\,s^{-1}}$, then

$$-\frac{vxI_n}{c^2} = \frac{10^{-4}}{9 \times 10^{16}} = 1.1 \times 10^{-21}\,\mathrm{C\,m^{-1}}$$

This confirms that in (8.22) the first term is about 10^{25} as large as the second term, but since λ_n is cancelled *perfectly* by λ_p, only the small $\boldsymbol{B}$ term remains.

Of course, the force between two current-carrying wires is appreciable only because the second wire has also, say, 10^{29} electrons per m^3. In fact the force of attraction between two long, thin, parallel wires each carrying 1 A in the same direction and places 1 m apart is exactly $2 \times 10^{-7}\,\mathrm{N\,m^{-1}}$, from the definition of the ampere in Chapter 4.

8.3 VECTOR POTENTIALS

In electrostatics it is often the case that to find the electric field $\boldsymbol{E}$ from a distribution of charges $\rho\,\mathrm{d}\tau$, it is easier to find first the electric scalar potential ϕ from an integral, (2.14),

$$\phi(1) = \frac{1}{4\pi\varepsilon_0} \int_{\substack{\text{all}\\ \text{space}}} \frac{\rho(2)\,d\tau_2}{r_{12}} \tag{8.24}$$

and then compute the electric field from (2.17),

$$\boldsymbol{E} = -\operatorname{grad}\phi \tag{8.25}$$

In electromagnetism a similar procedure for finding the magnetic field $\boldsymbol{B}$ from a distribution of moving charges is possible in terms of the magnetic vector potential, $\boldsymbol{A}$. Since $\boldsymbol{B}$ is always a divergence-free field, by (7.2), we can always write

$$\boldsymbol{B} = \operatorname{curl}\boldsymbol{A} \tag{8.26}$$

which makes

$$\operatorname{div}\boldsymbol{B} = \operatorname{div}\operatorname{curl}\boldsymbol{A} = \nabla\cdot(\nabla\times\boldsymbol{A}) = 0$$

8.3.1 Moving charge

We have seen that a charge q travelling at speed u along the x axis and passing through the origin at $t = 0$ (Fig. 8.5) produces electric and magnetic fields at P given by (8.16) (we have dropped the subscript 1 here). We shall now show that the vector potential for this moving charge is

$$\boldsymbol{A} = \frac{\mu_0 \gamma q u}{4\pi(r'^2)^{1/2}}\hat{\boldsymbol{\imath}} \tag{8.27}$$

where r' in the moving frame is the Lorentz transform of r in the laboratory frame and, by (8.13),

$$r'^2 = \gamma^2(x - ut)^2 + y^2 + z^2 \tag{8.28}$$

By definition

$$\boldsymbol{B} = \operatorname{curl}\boldsymbol{A} = \left(\frac{\partial A_z}{\partial y} - \frac{\partial A_y}{\partial z}\right)\hat{\mathbf{i}} + \left(\frac{\partial A_x}{\partial z} - \frac{\partial A_z}{\partial x}\right)\hat{\mathbf{j}} + \left(\frac{\partial A_y}{\partial x} - \frac{\partial A_x}{\partial y}\right)\hat{\mathbf{k}}$$

and so in this case

$$\boldsymbol{B} = \frac{\partial A_x}{\partial z}\hat{\mathbf{j}} - \frac{\partial A_x}{\partial y}\hat{\mathbf{k}}$$

$$= \frac{\mu_0 \gamma q u}{4\pi}\left\{-\frac{1}{2}\cdot\frac{2z}{(r'^2)^{3/2}}\hat{\mathbf{j}} + \frac{1}{2}\frac{2y}{(r'^2)^{3/2}}\hat{\mathbf{k}}\right\}$$

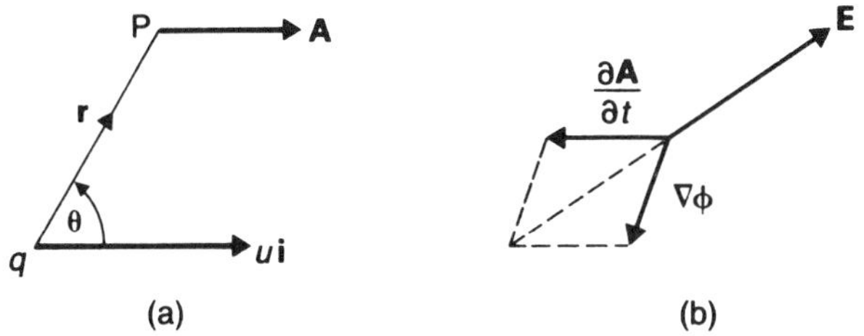

Fig. 8.7 (a) Vector potential and (b) electric field of a moving charge.

or
$$B = \frac{\mu_0 \gamma q u}{4\pi(r'^3)}(-z\hat{\mathbf{j}} + y\hat{\mathbf{k}})$$

exactly as found before in (8.16). Therefore the vector potential A of a moving charge (Fig. 8.7(a)) is parallel to the velocity vector of the charge and varies as r^{-1}.

When a charge is moving its electric field is no longer given by (8.25) but by

$$E = -\operatorname{grad}\phi - \frac{\partial A}{\partial t} \tag{8.29}$$

which, with (8.26) and the identity curl grad $\phi = 0$, is Maxwell's equation (7.3). We can see that this is correct for the charge in Fig. 8.7(a), where A is given by (8.27) and the scalar potential is

$$\phi = \frac{\gamma q}{4\pi\varepsilon_0 r'} \tag{8.30}$$

In full:

$$E = -\frac{\partial\phi}{\partial x}\hat{\mathbf{i}} - \frac{\partial\phi}{\partial y}\hat{\mathbf{j}} - \frac{\partial\phi}{\partial z}\hat{\mathbf{k}} - \frac{\partial A}{\partial t}$$

$$= \frac{\gamma q}{4\pi\varepsilon_0(r'^3)}\left\{[\gamma^2(x - ut)\hat{\mathbf{i}} + y\hat{\mathbf{j}} + z\hat{\mathbf{k}}] - \mu_0\varepsilon_0 u^2\gamma^2(x - ut)\hat{\mathbf{i}}\right\}$$

since $\partial(r'^2)/\partial t = -2\gamma^2 u(x - ut)$. Hence

$$E = \frac{\gamma q}{4\pi\varepsilon_0(r'^3)}r$$

since $\gamma^2 - \gamma^2(u^2/c^2) = \gamma^2(1 - \beta^2) = 1$, exactly as before in (8.16).

8.3.2 Four-vector equations

The potentials $\boldsymbol{A}$ and ϕ given in (8.26) and (8.29) are not unique but would still satisfy these equations if $\boldsymbol{A}$ was $(\boldsymbol{A} + \text{grad}\,\psi)$ and ϕ was $(\phi + \phi_0)$ for example. Lorentz showed that if we choose the gauge

$$\text{div}\,\boldsymbol{A} = -\frac{1}{c^2}\frac{\partial \phi}{\partial t} \tag{8.31}$$

then Maxwell's equations can be expressed in a particularly simple form. Putting (8.29) and (8.31) into the first Maxwell equation (7.1),

$$\text{div}\,\boldsymbol{E} = \rho/\varepsilon_0$$

we obtain

$$-\nabla^2\phi - \frac{\partial}{\partial t}(\text{div}\,\boldsymbol{A}) = \frac{\rho}{\varepsilon_0}$$

and so

$$\nabla^2\phi - \frac{1}{c^2}\frac{\partial^2\phi}{\partial t^2} = -\frac{\rho}{\varepsilon_0} \tag{8.32}$$

Similarly putting (8.26), (8.29) and (8.31) into the fourth Maxwell equation, (7.4),

$$\text{curl}\,\boldsymbol{B} = \mu_0\left(\boldsymbol{j} + \varepsilon_0\frac{\partial \boldsymbol{E}}{\partial t}\right)$$

we obtain

$$\text{grad div}\,\boldsymbol{A} - \nabla^2\boldsymbol{A} = \mu_0\boldsymbol{j} - \frac{1}{c^2}\frac{\partial}{\partial t}(\text{grad}\,\phi) - \frac{1}{c^2}\frac{\partial^2\boldsymbol{A}}{\partial t^2}$$

and so

$$\nabla^2\boldsymbol{A} - \frac{1}{c^2}\frac{\partial^2\boldsymbol{A}}{\partial t^2} = -\mu_0\boldsymbol{j} \tag{8.33}$$

In the four dimensions of special relativity, the Laplacian operator ∇^2 is replaced by the D'Alembertian operator

$$\Box = \nabla^2 - \frac{1}{c^2}\frac{\partial^2}{\partial t^2} \tag{8.34}$$

so that (8.32) and (8.33) can be written

$$\Box\phi = -\rho/\varepsilon_0 \quad \text{and} \quad \Box\boldsymbol{A} = -\mu_0\boldsymbol{j} \tag{8.35}$$

We have already seen (section 8.1) that

$$j_\nu \equiv (\boldsymbol{j}, ic\rho) \tag{8.36}$$

is a four-vector and so corresponds to the right-hand side of the four equations that form (8.35) divided by ε_0. The D'Alembertian, like the Laplacian, is the same for all coordinate systems, so the quantities A_x, A_y, A_z, ϕ must also form a four-potential:

$$A_\nu \equiv (\boldsymbol{A}, i\phi/c) \tag{8.37}$$

The simplicity of Maxwell's equations is now apparent, since we can write (8.33) in their invariant, relativistic form as:

$$\Box A_\nu = -\mu_0 j_\nu \tag{8.38}$$

Similarly, using the four-dimensional vector operator $\nabla_\nu \equiv (\nabla, i\partial/c\partial t)$ the Lorentz condition becomes

$$\nabla_\nu A_\nu = 0 \tag{8.39}$$

8.3.3 Biot–Savart law

Since the vector potential $\boldsymbol{A}$ and the scalar potential ϕ are the components of the four-potential A_ν, many of the problems solved in electrostatics from Poisson's equation $\nabla^2\phi = -\rho/\varepsilon_0$ can be similarly solved in magnetostatics from the equation

$$\nabla^2 \boldsymbol{A} = -\mu_0 \boldsymbol{j} \tag{8.40}$$

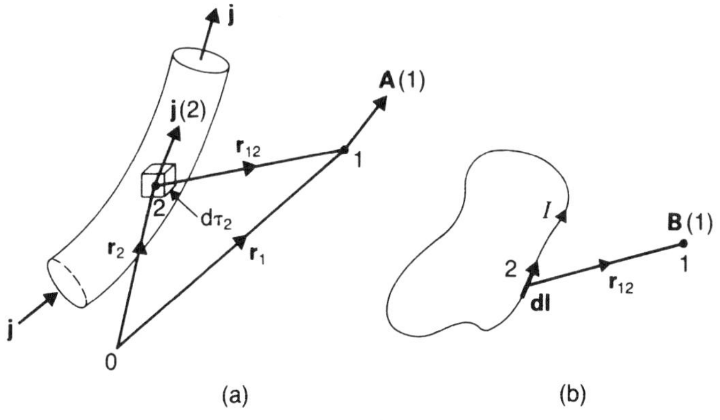

Fig. 8.8 (a) Vector potential $\boldsymbol{A}(1)$ for a distribution of current density $\boldsymbol{j}$. (b) Magnetic field $\boldsymbol{B}(1)$ due to current I.

Thus (8.24) for ϕ becomes

$$\boldsymbol{A}(1)=\frac{\mu_0}{4\pi}\int_{\substack{\text{all}\\\text{space}}}\frac{\boldsymbol{j}(2)\,\mathrm{d}\tau_2}{r_{12}} \tag{8.41}$$

for each component of the vector potential $\boldsymbol{A}(1)$ and the vector current density $\boldsymbol{j}(2)$, as in Fig. 8.8(a). Then we can find $\boldsymbol{B}(1)$ from

$$\boldsymbol{B}(1)=\operatorname{curl}\boldsymbol{A}(1)=\operatorname{curl}\left\{\frac{\mu_0}{4\pi}\int\frac{\boldsymbol{j}(2)\,\mathrm{d}\tau_2}{r_{12}}\right\}$$

where

$$r_{12}^2=(x_1-x_2)^2+(y_1-y_2)^2+(z_1-z_2)^2$$

In finding the derivatives of $\boldsymbol{A}(1)$ we operate only on the $(x_1y_1z_1)$ coordinates, so that

$$\begin{aligned}B_x&=\frac{\partial A_z}{\partial y_1}-\frac{\partial A_y}{\partial z_1}=\frac{\mu_0}{4\pi}\int\left\{j_z\frac{\partial}{\partial y_1}\left(\frac{1}{r_{12}}\right)-j_y\frac{\partial}{\partial z_1}\left(\frac{1}{r_{12}}\right)\right\}\mathrm{d}\tau_2\\&=\frac{\mu_0}{4\pi}\int\left\{\frac{-j_z(y_1-y_2)}{r_{12}^3}+\frac{j_y(z_1-z_2)}{r_{12}^3}\right\}\mathrm{d}\tau_2\end{aligned}$$

Here the integrand is just the x component of $(\boldsymbol{j}\times\boldsymbol{r}_{12})/r_{12}^3$ and, by symmetry, we therefore find

$$\boldsymbol{B}(1)=\frac{\mu_0}{4\pi}\int_{\substack{\text{all}\\\text{space}}}\frac{\boldsymbol{j}(2)\times\boldsymbol{r}_{12}}{r_{12}^3}\,\mathrm{d}\tau_2 \tag{8.42}$$

In many circuits the current is carried in wires whose diameters are very small compared with the other dimensions of the circuit. For thin wires the volume element $\mathrm{d}\tau=S\,\mathrm{d}l$ and the current density $\boldsymbol{j}$ is along $\mathbf{d}\boldsymbol{l}$ and uniform over S, so that

$$\boldsymbol{j}\,\mathrm{d}\tau=\boldsymbol{j}S\,\mathrm{d}l=I\,\mathbf{d}\boldsymbol{l} \tag{8.43}$$

where I is the current in the circuit. Then (8.42) becomes

$$\boldsymbol{B}(1)=\frac{\mu_0}{4\pi}\oint\frac{I\,\mathbf{d}\boldsymbol{l}\times\boldsymbol{r}_{12}}{r_{12}^3} \tag{8.44}$$

where the integration is taken all round the circuit (Fig. 8.8(b)). This is the law of Biot and Savart for steady currents, already expressed in (4.17).

8.4 ENERGY OF ELECTROMAGNETIC FIELD

We consider here the energy that arises from fixed charges electrostatically and from steady currents magnetostatically. Later we shall consider the energy density under dynamic conditions such as that carried by an electromagnetic wave.

8.4.1 Electrostatic energy

The potential energy of an infinitesimal volume of charge $\rho\,\mathrm{d}\tau$ is the product $\rho\phi\,\mathrm{d}\tau$, where ϕ is the potential due to any other charges (Fig. 8.9). To find the total electrostatic energy U we must integrate this over the charge distribution, remembering that such an integral would count all the pairs $\rho_i\phi_i\,\mathrm{d}\tau$ twice. Therefore,

$$U = \frac{1}{2}\int_\tau \rho\phi\,\mathrm{d}\tau \tag{8.45}$$

This expression is adequate for a distribution of fixed charges, but if we integrate it using Poisson's equation $\nabla^2\phi = -\rho/\varepsilon_0$, we can obtain a more general result in terms of the electric field $\boldsymbol{E}$. We have

$$U = -\frac{\varepsilon_0}{2}\int_\tau \phi\nabla^2\phi\,\mathrm{d}\tau$$

or

$$U = -\frac{\varepsilon_0}{2}\int_\tau (\phi\nabla\cdot\nabla\phi)\,\mathrm{d}\tau$$

Using the vector identity for div $\boldsymbol{\Omega A}$, where $\boldsymbol{\Omega}$ and $\boldsymbol{A}$ are arbitrary scalar and vector functions (Appendix E, equation E1), this

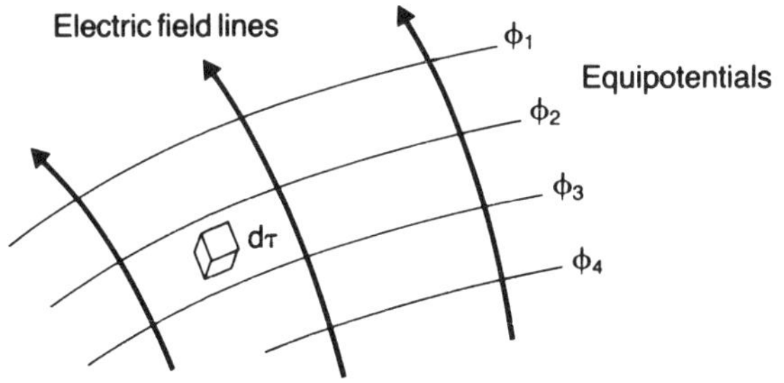

Fig. 8.9 Equipotentials in an electric field.

becomes:

$$U = -\frac{\varepsilon_0}{2}\int_\tau \nabla\cdot(\phi\nabla\phi)\,\mathrm{d}\tau + \frac{\varepsilon_0}{2}\int_\tau \nabla\phi\cdot\nabla\phi\,\mathrm{d}\tau$$

Applying Gauss's divergence theorem (Appendix E, equation E19) to the first integral

$$U = -\frac{\varepsilon_0}{2}\int_S (\phi\nabla\phi)\cdot\mathbf{dS} + \frac{\varepsilon_0}{2}\int_\tau \boldsymbol{E}\cdot\boldsymbol{E}\,\mathrm{d}\tau$$

For a finite volume τ of charges, the surface S can be made as large as we wish. The integrand is the product $\phi\nabla\phi$, which must decrease at least as fast as $1/r \times 1/r^2 = 1/r^3$, whereas the surface area of S will increase only as r^2. Thus this contribution to the potential energy is negligible and

$$U = \frac{\varepsilon_0}{2}\int_\tau \boldsymbol{E}\cdot\boldsymbol{E}\,\mathrm{d}\tau \qquad (8.46)$$

An energy density $u = U/\tau$ in the electrostatic field (Fig. 8.9) of $\frac{1}{2}\varepsilon_0 E^2$ would produce the same total energy U and this interpretation of the electric energy is particularly useful for electromagnetism. On the other hand, it is only for fixed charges that one can write (8.45).

8.4.2 Magnetostatic energy

In establishing a distribution of steady currents there is an initial transient period when the currents and their associated fields are brought from zero to their final values. During this period there are time-dependent fields which induce electromotive forces, and so the total magnetic energy of steady currents must include work done in establishing those currents. For a single circuit carrying current I, if the magnetic flux Φ through it changes, then an e.m.f. $\mathscr{E}$ is induced around it. In order to keep the current I constant, the batteries must do work at the rate

$$\frac{\mathrm{d}U}{\mathrm{d}t} = -I\mathscr{E} = \frac{I\,\mathrm{d}\Phi}{\mathrm{d}t}$$

from Faraday's law. Such a circuit can be considered to be composed of a large number of current loops (Fig. 8.10), each of which carries the current I around an infinitesimal area dS with normal $\mathbf{dS}$. An

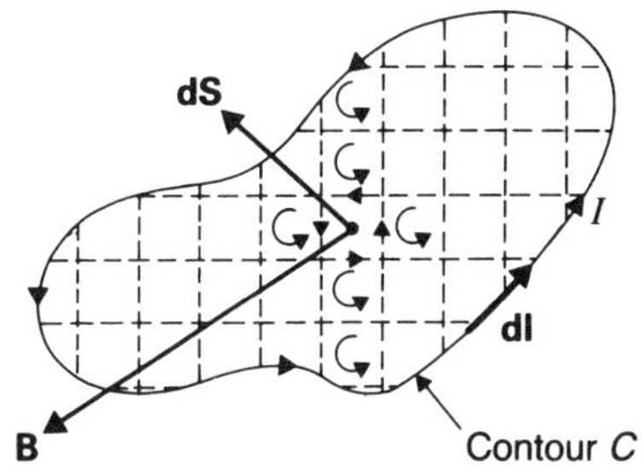

Fig. 8.10 A circuit carrying a current I is equivalent to a large number of current loops of magnetic dipole moment $I\,\mathbf{d}\boldsymbol{S}$ and this can be used to find the magnetic energy of the circuit in a magnetic field $\boldsymbol{B}$.

increment of work $\mathrm{d}U$ done against the induced e.m.f. in the loop is given by the change in the magnetic field $\mathbf{d}\boldsymbol{B}$ through the loop, i.e.,

$$\mathrm{d}U = I\,\mathbf{d}\boldsymbol{S}\cdot\mathbf{d}\boldsymbol{B}$$

The total energy in establishing the magnetic field $\boldsymbol{B}$ of all the loops is then

$$U = I\int_S \boldsymbol{B}\cdot\mathbf{d}\boldsymbol{S} \tag{8.47}$$

and expressing $\boldsymbol{B}$ in terms of the vector potential

$$U = I\int_S (\mathrm{curl}\,\boldsymbol{A})\cdot\mathbf{d}\boldsymbol{S}$$

By Stokes's theorem (Appendix E, equation E20) this integral over the total surface S of the circuit can be expressed in terms of its contour C by

$$U = I\oint_C \boldsymbol{A}\cdot\mathbf{d}\boldsymbol{l} \tag{8.48}$$

The thin wires of the circuit can be replaced by current filaments, such that

$$I\,\mathbf{d}\boldsymbol{l} = \boldsymbol{j}\,\mathrm{d}\tau \tag{8.43}$$

and for each pair of such circuits the magnetic energy is given by (8.48), where the integral is over the volume of the current filaments of one circuit and the vector potential is due to the other circuit. For the total energy if we take the sum over all filaments we should be counting each filament twice so that, like (8.45), we obtain

$$U = \frac{1}{2}\int_\tau \boldsymbol{j}\cdot\boldsymbol{A}\,\mathrm{d}\tau \tag{8.49}$$

This expression is adequate for a distribution of steady currents, but if we interpret it using Maxwell's fourth equation, (7.4)

$$\operatorname{curl}\boldsymbol{B} = \mu_0\boldsymbol{j} \tag{8.50}$$

we can obtain a more general result in terms of the magnetic field $\boldsymbol{B}$. Hence

$$U = \frac{1}{2}\mu_0\int_\tau (\operatorname{curl}\boldsymbol{B})\cdot\boldsymbol{A}\,\mathrm{d}\tau$$

Using the vector identity for div ($\boldsymbol{A} \times \boldsymbol{B}$), where $\boldsymbol{A}$ and $\boldsymbol{B}$ are arbitrary vectors (Appendix E, equation E2), this becomes

$$U = \frac{1}{2}\mu_0\int_\tau \{\boldsymbol{B}\cdot\operatorname{curl}\boldsymbol{A} - \operatorname{div}(\boldsymbol{A}\times\boldsymbol{B})\}\,\mathrm{d}\tau$$

Applying Gauss's divergence theorem (Appendix E, equation E19) to the second term, this becomes

$$U = \frac{1}{2}\mu_0\int_\tau \boldsymbol{B}\cdot\boldsymbol{B}\,\mathrm{d}\tau - \frac{1}{2}\mu_0\int_S (\boldsymbol{A}\times\boldsymbol{B})\cdot\mathbf{d}\boldsymbol{S}$$

For a finite system of currents, the surface S can be made as large as we wish. The integral over a surface a long way from the system would vanish, since $\boldsymbol{A} \times \boldsymbol{B}$ falls off at least as fast as r^{-3}, while S increases only as r^2. Hence the total magnetic energy of a system of steady currents in conductors is

$$U = \frac{\mu_0}{2}\int_\tau \boldsymbol{B}\cdot\boldsymbol{B}\,\mathrm{d}\tau \tag{8.51}$$

As with the electrostatic field, an energy density $u = U/\tau$ in the magnetostatic field of $\frac{1}{2}\mu_0 B^2$ would produce the same total energy U and this interpretation is particularly useful in electromagnetism. On the other hand it is only for a system of steady currents that one can use (8.49).

8.5 RETARDED POTENTIALS

We have shown that Maxwell's four equations for the $\boldsymbol{E}$ and $\boldsymbol{B}$ fields, (7.1) to (7.4), are equivalent to one four-vector equation (8.38)

$$\Box A_\nu = -\mu_0 j_\nu$$

provided we choose the Lorentz gauge for A_ν, (8.39), i.e.

$$\nabla_\nu A_\nu = 0$$

Here A_ν was the four-vector $(\boldsymbol{A}, i\phi/c)$, so that (8.38) is a set of three equations for the components of $\boldsymbol{A}$ and an equation for ϕ, (8.33) and (8.32), respectively:

$$\nabla^2 \boldsymbol{A} - \frac{1}{c^2}\frac{\partial^2 \boldsymbol{A}}{\partial t^2} = -\mu_0 \boldsymbol{j}$$

$$\nabla^2 \phi - \frac{1}{c^2}\frac{\partial^2 \phi}{\partial t^2} = -\rho/\varepsilon_0$$

For a distribution of charges and currents (Fig. 8.11) we can integrate these *inhomogeneous wave equations* in a similar way to the integration of Poisson's equation, $\nabla^2\phi = -\rho/\varepsilon_0$, which yields the familiar scalar potential given in (8.24), i.e.

$$\phi(1) = \frac{1}{4\pi\varepsilon_0}\int \frac{\rho(2)}{r_{12}}\,\mathrm{d}\tau_2$$

The difference, now that we have moving charges, from the electrostatic result is that the potentials at the fixed field points $[1, t] = (x_1 y_1 z_1 t)$ are due to the charges and currents at the source points at the earlier time $(t - r_{12}/c)$ to allow for the fields to propagate at the finite speed c.

The solutions of (8.33) and (8.32) are, therefore, the retarded vector

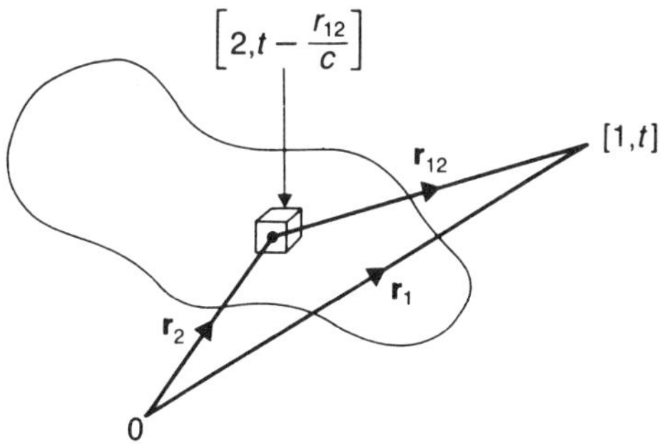

Fig. 8.11 The retarded potentials at point $[1, t]$ are calculated from integrals of the charge density and current density at source points $[2, t - r_{12}/c]$ within a source distribution.

and scalar potentials:

$$\boldsymbol{A}(1,t) = \frac{\mu_0}{4\pi}\int \frac{\boldsymbol{j}(2, t - r_{12}/c)}{r_{12}} \mathrm{d}\tau_2 \tag{8.52}$$

$$\phi(1,t) = \frac{1}{4\pi\varepsilon_0}\int \frac{\rho(2, t - r_{12}/c)}{r_{12}} \mathrm{d}\tau_2 \tag{8.53}$$

These potentials replace the static potentials of (8.41) and (8.24) whenever we are not dealing with the separate electrostatic ($\partial \boldsymbol{E}/\partial t = 0$) or magnetostatic ($\partial \boldsymbol{B}/\partial t = 0$) solutions of Maxwell's equations. They show clearly the relativistic nature of the electromagnetic field and the speed of propagation of electromagnetism in space. They are expressed in the four-vector notation by

$$A_\nu(1,t) = \frac{\mu_0}{4\pi}\int \frac{j_\nu(2, t - r_{12}/c)}{r_{12}} \mathrm{d}\tau_2 \tag{8.54}$$

9

Electromagnetic waves in space

Electromagnetic fields propagate in space, in dielectrics and, to a limited extent, in plasmas and conductors. The basic properties of electromagnetic waves are revealed by studying their propagation in space, so we begin by solving Maxwell's equations in free space for their wave solutions and then determine the flow of energy in these waves.

9.1 WAVE EQUATIONS

In empty space there can be no electric charges or currents, so Maxwell's equations, (7.1) to (7.4) respectively, became

$$\operatorname{div} \boldsymbol{E} = 0$$

$$\operatorname{div} \boldsymbol{B} = 0$$

$$\operatorname{curl} \boldsymbol{E} = -\,\partial \boldsymbol{B}/\partial t$$

$$\operatorname{curl} \boldsymbol{B} = \mu_0 \varepsilon_0\, \partial \boldsymbol{E}/\partial t$$

By taking the curl of (7.8) and (7.9) and using the vector identity for curl curl $\boldsymbol{A}$, where $\boldsymbol{A}$ is any vector (Appendix E, equation E6) we obtain

$$\operatorname{grad} \operatorname{div} \boldsymbol{E} - \nabla^2 \boldsymbol{E} = -\operatorname{curl}(\partial \boldsymbol{B}/\partial t) = -\frac{\partial}{\partial t}(\operatorname{curl} \boldsymbol{B})$$

$$\operatorname{grad} \operatorname{div} \boldsymbol{B} - \nabla^2 \boldsymbol{B} = \mu_0 \varepsilon_0 \operatorname{curl}(\partial \boldsymbol{E}/\partial t) = \mu_0 \varepsilon_0 \frac{\partial}{\partial t}(\operatorname{curl} \boldsymbol{E})$$

since the space operators are independent of the time coordinates. Remembering (7.6), (7.7) and re-using (7.8) and (7.9), these become

$$\nabla^2 \boldsymbol{E} = \mu_0 \varepsilon_0 \partial^2 \boldsymbol{E}/\partial t^2 \tag{9.1}$$

$$\nabla^2 \boldsymbol{B} = \mu_0 \varepsilon_0 \partial^2 \boldsymbol{B}/\partial t^2 \tag{9.2}$$

The electric constant ε_0 is defined as $1/(\mu_0 c^2)$, so that (9.1) and (9.2) are

$$\nabla^2 \boldsymbol{E} - \frac{1}{c^2}\frac{\partial^2 \boldsymbol{E}}{\partial t^2} = 0 \tag{9.3}$$

$$\nabla^2 \boldsymbol{B} - \frac{1}{c^2}\frac{\partial^2 \boldsymbol{B}}{\partial t^2} = 0 \tag{9.4}$$

We thus have the remarkable result that, in space, the electric field $\boldsymbol{E}(\boldsymbol{r}, t)$, the magnetic field $\boldsymbol{B}(\boldsymbol{r}, t)$, the scalar potential $\phi(\boldsymbol{r}, t)$ and the vector potential $\boldsymbol{A}(\boldsymbol{r}, t)$, from (9.2), (9.4), (8.32) and (8.33), all satisfy the same basic, four-dimensional wave equation

$$\nabla^2 \psi = \frac{1}{c^2}\frac{\partial^2 \psi}{\partial t^2} \tag{9.5}$$

This equation is well known and its general solution is a superposition of an infinite set of one-dimensional waves travelling in all possible directions. We can therefore obtain the basic information we need by studying the equation for one spatial dimension and the time. That is

$$\frac{\partial^2 \psi}{\partial z^2} = \frac{1}{c^2}\frac{\partial^2 \psi}{\partial t^2} \tag{9.6}$$

Let us suppose that at $t = 0$ the solution is an arbitrary function, $f(z)$, as shown in Fig. 9.1. It travels along $0z$ at speed c, so that at

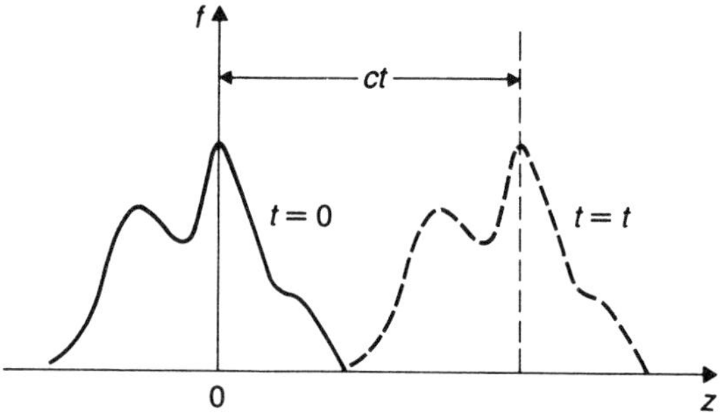

Fig. 9.1 A plane wave $\psi(z, t)$ travelling along $0z$ at speed c in the time t.

time t it has travelled a distance ct and in space it will not be distorted. The solution is therefore of the form

$$\psi = f(z - ct) \tag{9.7}$$

which the reader can readily see satisfies (9.6) by partial differentiation. It is also possible for the wave to be travelling at speed c in the opposite direction, $-z$. A general solution of (9.6) is therefore

$$\psi = f(z - ct) + g(z + ct) \tag{9.8}$$

There is no need to introduce a particular frequency into this general solution, as there is no dispersion for electromagnetic waves in space. All such waves, from the highest-frequency cosmic rays (see electromagnetic spectrum) to the lowest-frequency radio waves, travel in space at exactly the same speed, c. This has been verified over many years in numerous precision experiments, notably those with monochromatic light and microwaves. The accepted value for c from these experiments, to an accuracy of 4 parts in 10^9, is

$$c = 299\,792\,458 \text{ m s}^{-1} \tag{9.9}$$

However, the SI base units of length and time are now defined in terms of the wavelength λ of a transition of the ^{86}Kr atom ($1 \text{ m} = 1\,650\,763.73\,\lambda$) and the frequency ν of a transition of the ^{133}Cs atom ($1 \text{ s} = 9\,192\,631\,770\,\nu^{-1}$). Since the radiation from each of these transitions travels at speed c, a measurement of c is now directly related to these transitions. For problems it is usually acceptable to take $c = 3.00 \times 10^8 \text{ m s}^{-1}$.

9.2 PLANE WAVES

The simplest waves are plane waves, i.e. waves where the fields are constant in a plane (say, xy) at an instant in time for a wave propagating along an axis normal to the plane (say, z). The electric field, for example, in a plane electromagnetic wave is given by the real part of

$$\boldsymbol{E} = \boldsymbol{E}_0 \exp \mathrm{i}(\omega t - kz) \tag{9.10}$$

where the time-period $T = 2\pi/\omega$, the wavelength $\lambda = 2\pi/k$ and the phase velocity $\omega/k = c$ (Fig. 9.2). (It is conventional in electromagnetism and optics to write a wavefunction in this way with $+\mathrm{i}\omega t$, although in quantum mechanics and solid state physics the opposite convention of $-\mathrm{i}\omega t$ is usual.)

In our plane wave at a fixed time (Fig. 9.2(b)) the field vector $\boldsymbol{E}$

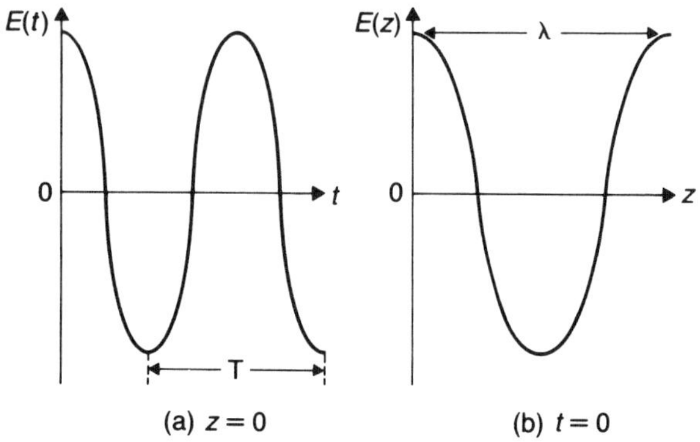

Fig. 9.2 A plane wave $E(z, t)$ has a time period T when observed (a) at $z = 0$ and a wavelength λ measured (b) at $t = 0$ along $0z$.

is constant in the xy plane, so that

$$\frac{\partial E_x}{\partial x} = \frac{\partial E_y}{\partial y} = 0 \tag{9.11}$$

But the wave must satisfy Maxwell's first equation (7.6)

$$\frac{\partial E_x}{\partial x} + \frac{\partial E_y}{\partial y} + \frac{\partial E_z}{\partial z} = 0 \tag{9.12}$$

Combining (9.11) and (9.12), we have

$$\frac{\partial E_z}{\partial z} = 0 \tag{9.13}$$

or E_z is constant. However, from (9.10),

$$\frac{\partial \boldsymbol{E}}{\partial z} = -k\boldsymbol{E} \tag{9.14}$$

and the only value of E_z that will satisfy both (9.13) and (9.14) is

$$E_z = 0$$

Therefore $\boldsymbol{E}_0$ is a vector in the xy plane normal to the direction of propagation $0z$.

In a similar way the magnetic field in a plane electromagnetic wave is given by the real part of

$$\boldsymbol{B} = \boldsymbol{B}_0 \exp \mathrm{i}(\omega t - kz) \tag{9.15}$$

Since $\boldsymbol{B}$ must satisfy Maxwell's second equation (7.7),

$$\operatorname{div} \boldsymbol{B} = 0$$

a similar argument shows that $B_z = 0$ and therefore $\boldsymbol{B}_0$ is also a vector in the xy plane. Of course, $\boldsymbol{E}$ and $\boldsymbol{B}$ are related in Maxwell's third and fourth equations, so there must be a relationship between $\boldsymbol{E}_0$ and $\boldsymbol{B}_0$ in an electromagnetic wave. To find this, let

$$\boldsymbol{E}_0 = E_{0x}\hat{\mathbf{i}} + E_{0y}\hat{\mathbf{j}} \tag{9.16}$$

where E_{0x}, E_{0y} are constants and $\hat{\mathbf{i}}, \hat{\mathbf{j}}$ are unit vectors. Substituting (9.10) and (9.16) in Maxwell's third equation (7.8) we have

$$\operatorname{curl}\{(E_{0x}\hat{\mathbf{i}} + E_{0y}\hat{\mathbf{j}})\exp \mathrm{i}(\omega t - kz)\} = -\frac{\partial \boldsymbol{B}}{\partial t} \tag{9.17}$$

The only components of curl $\boldsymbol{E}$ that are finite are $\partial E_x/\partial z$ and $\partial E_y/\partial z$, so that (9.17) becomes

$$(-\mathrm{i}kE_{0x}\hat{\mathbf{j}} + \mathrm{i}kE_{0y}\hat{\mathbf{i}})\exp \mathrm{i}(\omega t - kz) = -\partial \boldsymbol{B}/\partial t$$

If we now integrate to find $\boldsymbol{B}$ and take it to be entirely oscillatory, we divide by iω and get:

$$\frac{k}{\omega}E_{0x}\hat{\mathbf{j}} - \frac{k}{\omega}E_{0y}\hat{\mathbf{i}} = \boldsymbol{B}_0$$

However, $c = \omega/k$ is the phase velocity and so the vector $\boldsymbol{B}$ is just

$$\boldsymbol{B} = \frac{1}{c}(\hat{\mathbf{k}} \times \boldsymbol{E}) \tag{9.18}$$

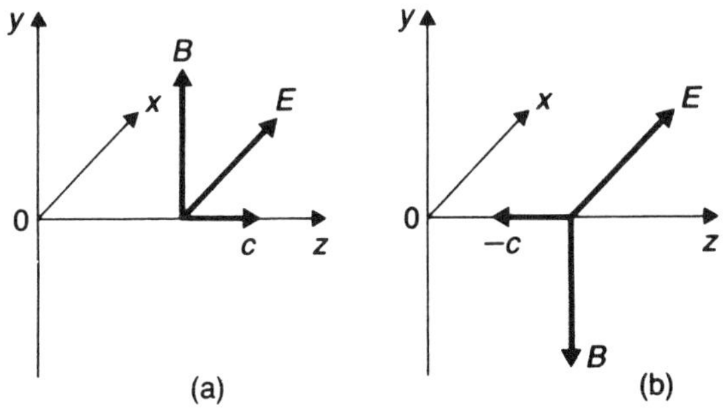

Fig. 9.3 A plane electromagnetic wave travelling (a) in the $+z$ direction and (b) in the $-z$ direction.

This shows that $\boldsymbol{B}$ is normal to both $\boldsymbol{E}$ and the direction of propagation ($\hat{\mathbf{k}}$) and similarly for $\boldsymbol{E}$. Electromagnetic waves are therefore *transverse waves.* The relative directions of $\boldsymbol{E}$ and $\boldsymbol{B}$ are shown in Fig. 9.3 for waves travelling along $+z$ and $-z$. Of course, the amplitude of $\boldsymbol{B}$ is very small; for example, $1\,\mathrm{kV\,m^{-1}}$ for E corresponds to only $3.3\,\mu\mathrm{T}$ for B. (Note that $\boldsymbol{k}$ is the wave vector of the wave travelling in the direction $\hat{\mathbf{k}}$.)

9.2.1 Polarization

The plane wave described by (9.16) is said to be *linearly* polarized, because the electric field exists in one transverse direction only as it propagates (Fig. 9.4(a)), given by

$$\tan\theta = \frac{E_{0y}}{E_{0x}} \tag{9.19}$$

In this case the x and y components of $\boldsymbol{E}_0$ can be of different magnitudes but they are always in phase and so θ is a constant.

A plane wave composed of x and y components that have a constant phase difference, as well as different amplitudes, is *elliptically* polarized; a special case is when the amplitudes are the same and the wave is then said to be *circularly* polarized. In general, then

$$\boldsymbol{E} = E_{0x}\hat{\mathbf{i}}\exp\mathrm{i}(\omega t - kz + \phi_x) + E_{0y}\hat{\mathbf{j}}\exp\mathrm{i}(\omega t - kz + \phi_y)$$

and if $\phi_y - \phi_x = \pi/2$, the real part is

$$\boldsymbol{E} = E_{0x}\hat{\mathbf{i}}\cos(\omega t - kz + \phi_x) - E_{0y}\hat{\mathbf{j}}\sin(\omega t - kz + \phi_x)$$

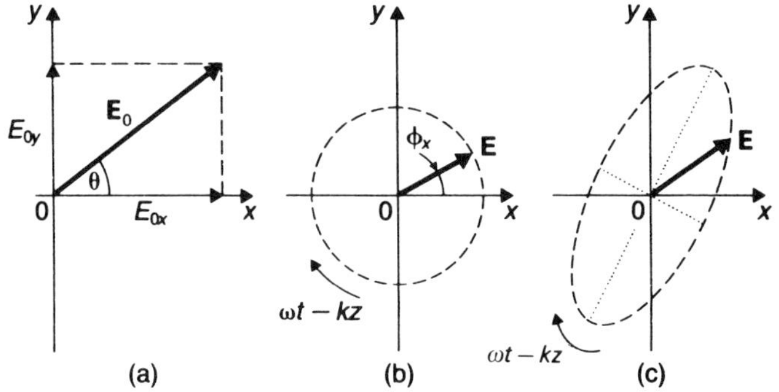

Fig. 9.4 The electric vector in a plane wave that is (a) linearly polarized, (b) circularly polarized, and (c) elliptically polarized.

Thus $\boldsymbol{E}$ rotates in the xy plane with direction,

$$\tan(\omega t - kz + \phi_x) = -\frac{E_{0y}}{E_{0x}} \tag{9.20}$$

which traces out a circle when $E_{0x} = E_{0y}$ (Fig. 9.4(b)) and an ellipse when $E_{0x} \neq E_{0y}$ (Fig. 9.4(c)).

The direction of rotation is defined as *right-handed* when, on looking into the wave along the direction of propagation, the electric vector is rotating counter-clockwise. The waves shown in Fig. 9.4(b) and (c) are thus *left-handed* polarizations. These definitions are the ones used in modern optics and in particle physics, the right-handed photon having positive helicity and spin vector in the direction of motion. (In classical optics the opposite convention was used.)

9.3 SPHERICAL WAVES

Plane waves are the simplest to describe mathematically and will be used extensively to study propagation in dielectrics and conductors in later chapters. In free space they propagate without any decrease in amplitude, unlike the waves from a point source or dipole, which spread out like the ripples from a stone thrown into a lake. In three dimensions such waves are spherical waves and it is instructive to rewrite the wave equation (9.5) in spherical polar coordinates and consider its spherically symmetric solution.

The Laplacian in spherical polars with spherical symmetry is (Appendix E, equation E14)

$$\nabla^2\psi = \frac{1}{r^2}\frac{\partial}{\partial r}\left(r^2\frac{\partial\psi}{\partial r}\right) \tag{9.21}$$

or

$$\nabla^2\psi = \frac{1}{r^2}\left\{2r\frac{\partial\psi}{\partial r} + r^2\frac{\partial^2\psi}{\partial r^2}\right\}$$

It is easily seen that this can be written as

$$\nabla^2\psi = \frac{1}{r}\left\{\frac{\partial^2}{\partial r^2}(r\psi)\right\}$$

so that (9.5) becomes

$$\frac{1}{r}\frac{\partial^2}{\partial r^2}(r\psi) = \frac{1}{c^2}\frac{\partial^2}{\partial t^2}(\psi) \tag{9.22}$$

or

$$\frac{\partial^2}{\partial r^2}(r\psi) = \frac{1}{c^2}\frac{\partial^2}{\partial t^2}(r\psi)$$

The solution is therefore of the same form as (9.7),

$$r\psi(r,t) = f(r - ct)$$

or

$$\psi(r,t) = \frac{1}{r}f(r - ct) \tag{9.23}$$

Thus a spherical wave decays in amplitude as $1/r$ (Fig. 9.5), unlike the constant amplitude plane wave (Fig. 9.1). There are two other ways in which these waves differ. First, it is obvious that the solution (9.23) cannot apply at the origin, $r = 0$, where the amplitude would be infinite. We shall see in Chapter 13, when we discuss the generation of electromagnetic waves, that the spherical wave solution is not valid when $r \ll \lambda$, the wavelength of the wave. Secondly, the solution which we found for the plane wave travelling in the opposite direction,

$$\psi(r,t) = \frac{1}{r}g(r + ct) \tag{9.24}$$

also satisfies (9.22), but implies a spherical wave collapsing to a source point. This is a valid solution of Maxwell's equations for waves from a spherical reflector but, like the outward solution (9.23), the inward solution (9.24) is not valid near the origin when $r \ll \lambda$.

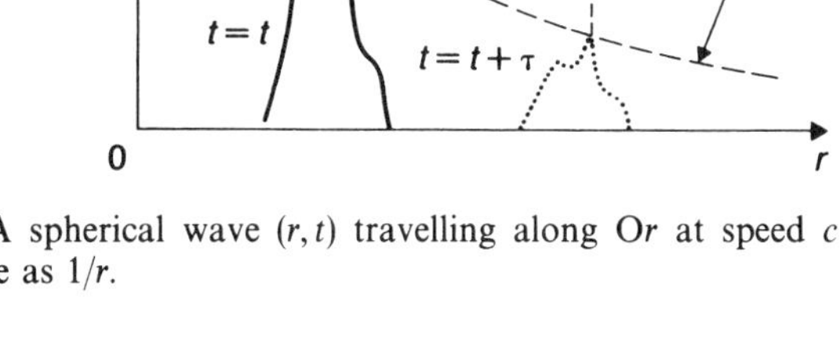

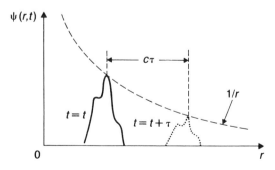

Fig. 9.5 A spherical wave (r, t) travelling along Or at speed c decays in amplitude as $1/r$.

9.4 ENERGY DENSITY AND ENERGY FLOW

An important principle in electromagnetism is the conservation of electric charge. It is recognized as of similar validity to the conservation of energy. In this section we consider the application of these conservation principles to the production and flow of energy in electromagnetic waves in space. We recall (Chapter 7) that, for a small volume $\mathrm{d}\tau$, we obtained a local conservation of electric charge, the *equation of continuity* (7.29)

$$\operatorname{div} \boldsymbol{j} = -\frac{\partial \rho}{\partial t}$$

This equation shows that the flux of charge density ρ flowing out of the volume $\mathrm{d}\tau$ is just equal to the rate of loss of charge density inside that volume. A similar conservation principle must apply to the energy flow into and out of a small volume, but now the charges are accelerated by an electric field $\boldsymbol{E}$ in order to produce an electromagnetic energy flux $\mathscr{S}$.

The rate of doing work on an electric charge q moving at velocity $\boldsymbol{v}$ is, by the Lorentz force law equation (8.10),

$$\boldsymbol{F} \cdot \boldsymbol{v} = q\boldsymbol{E} \cdot \boldsymbol{v}$$

or for a volume $\mathrm{d}\tau$

$$q\boldsymbol{E} \cdot \boldsymbol{v} = \rho \, \mathrm{d}\tau \, \boldsymbol{E} \cdot \boldsymbol{v} = \boldsymbol{E} \cdot \boldsymbol{j} \, \mathrm{d}\tau$$

If the energy density in space is u, then the rate of loss of energy density from any volume V must equal the flow of energy out of the surface S of that volume plus the rate of doing work on the moving charges in V that generates the energy flow, or

$$-\frac{\partial}{\partial t}\left(\int_V u \, \mathrm{d}\tau\right) = \int_S \mathscr{S} \cdot \mathbf{dS} + \int_V \boldsymbol{E} \cdot \boldsymbol{j} \, \mathrm{d}\tau$$

Applying Gauss's divergence theorem to the flux of $\mathscr{S}$, we obtain a similar equation to the equation of continuity, that is

$$-\frac{\partial u}{\partial t} = \operatorname{div} \mathscr{S} + \boldsymbol{E} \cdot \boldsymbol{j} \tag{9.25}$$

the *equation of conservation of energy flow*.

In 1884 Poynting solved this equation, using Maxwell's equations, to find the flux $\mathscr{S}$, which is now known as the Poynting vector, and the energy density u. We want to eliminate $\boldsymbol{j}$, and find $\mathscr{S}$ and u in terms of the electric $\boldsymbol{E}$ and magnetic $\boldsymbol{B}$ fields. Substituting for $\boldsymbol{j}$ from (7.4), we obtain

$$\boldsymbol{E}\cdot\boldsymbol{j} = \boldsymbol{E}\cdot\left(\operatorname{curl}\frac{\boldsymbol{B}}{\mu_0}\right) - \varepsilon_0\boldsymbol{E}\cdot\frac{\partial\boldsymbol{E}}{\partial t} \tag{9.26}$$

We recognize the second term as $\partial u/\partial t$ for the electric vector, from (8.46), and expect the first term to contain $\partial u/\partial t$ for the magnetic vector, from (8.51). Using the vector identity for $\operatorname{div}(\boldsymbol{A}\times\boldsymbol{B})$ from Appendix E, equation E2, the first term becomes

$$\boldsymbol{E}\cdot\left(\operatorname{curl}\frac{\boldsymbol{B}}{\mu_0}\right) = \operatorname{div}\left(\frac{\boldsymbol{B}}{\mu_0}\times\boldsymbol{E}\right) + \frac{\boldsymbol{B}}{\mu_0}\cdot(\operatorname{curl}\boldsymbol{E}) \tag{9.27}$$

Now, using Maxwell's equation (7.3), we see that the last term is

$$\frac{\boldsymbol{B}}{\mu_0}\cdot\left(-\frac{\partial\boldsymbol{B}}{\partial t}\right) = -\frac{\partial}{\partial t}\left(\frac{\boldsymbol{B}\cdot\boldsymbol{B}}{2\mu_0}\right) \tag{9.28}$$

and substituting from (9.27) and (9.28) in (9.26), we have

$$\boldsymbol{E}\cdot\boldsymbol{j} = \operatorname{div}\left(\frac{\boldsymbol{B}}{\mu_0}\times\boldsymbol{E}\right) - \frac{\partial}{\partial t}\left\{\frac{\boldsymbol{B}\cdot\boldsymbol{B}}{2\mu_0} + \frac{\varepsilon_0\boldsymbol{E}\cdot\boldsymbol{E}}{2}\right\}$$

Comparing this equation with (9.25), we find Poynting's solutions for $\mathscr{S}$ and u for energy flow in free space

$$\mathscr{S} = \boldsymbol{E}\times\frac{\boldsymbol{B}}{\mu_0} \equiv \boldsymbol{E}\times\boldsymbol{H} \tag{9.29}$$

$$u = \frac{1}{2}\left\{\frac{\boldsymbol{B}}{\mu_0}\cdot\boldsymbol{B} + \varepsilon_0\boldsymbol{E}\cdot\boldsymbol{E}\right\} \tag{9.30}$$

9.4.1 Poynting vector

We have seen that a plane electromagnetic wave in space with its electric vector along $\hat{\mathbf{i}}$ and its magnetic vector along $\hat{\mathbf{j}}$ (Fig. 9.3(a)) can be described by

$$\boldsymbol{E} = E_{0x}\exp\mathrm{i}(\omega t - kz)\hat{\mathbf{i}} \tag{9.31}$$

and

$$\boldsymbol{B} = \frac{1}{c}(\hat{\mathbf{k}}\times\boldsymbol{E}) = \frac{E_{0x}}{c}\exp\mathrm{i}(\omega t - kz)\hat{\mathbf{j}} \tag{9.32}$$

For this wave the Poynting vector is

$$\mathscr{S} = \boldsymbol{E} \times \frac{\boldsymbol{B}}{\mu_0} = \frac{\boldsymbol{E} \times (\hat{\mathbf{k}} \times \boldsymbol{E})}{c\mu_0} = \varepsilon_0 c E^2 \hat{\mathbf{k}} \tag{9.33}$$

This is the instantaneous rate of flow of energy across unit area, which is normally averaged over time to produce the intensity of the waves. So,

$$\langle \mathscr{S} \rangle = \varepsilon_0 c \langle E^2 \rangle \tag{9.34}$$

For example, a radio wave with $E = 0.1\ \mathrm{V\,m^{-1}}$ has an intensity of $10^{-2}/(4\pi \times 10^{-7} \times 3 \times 10^8) = 30\ \mu\mathrm{W\,m^{-2}}$.

The intensity can also be computed from the fluctuating fields through the energy density u (9.30). The plane wave of (9.31) and (9.32) is shown diagrammatically in Fig. 9.6(a), $\boldsymbol{E}$ oscillating in the xz plane and $\boldsymbol{B}$ in the yz plane. Although the mean values of $\boldsymbol{E}$ and $\boldsymbol{B}$ are zero, the average of the energy density

$$u = \frac{1}{2}\frac{E^2}{\mu_0 c^2} + \frac{1}{2}\varepsilon_0 E^2 \tag{9.35}$$

is not zero but

$$\langle u \rangle = \varepsilon_0 \langle E^2 \rangle$$

since $\varepsilon_0 = 1/(\mu_0 c^2)$ and the electric and magnetic energies are equal for an electromagnetic plane wave in space. The wave is travelling at speed c and so, from Fig. 9.6(b), the mean energy flow is just the

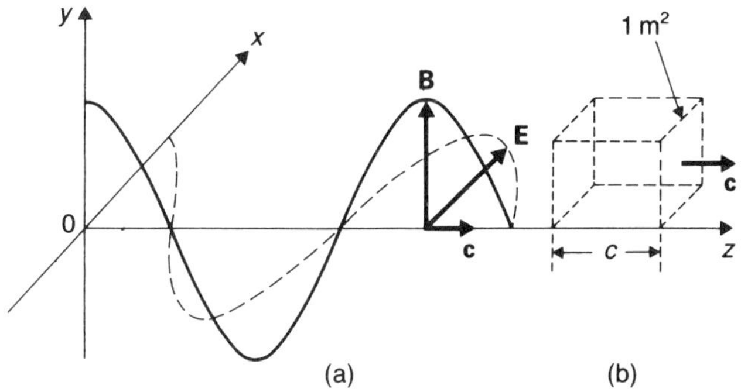

Fig. 9.6 (a) A plane electromagnetic wave travelling along Oz with electric vector in the xz plane, and magnetic vector in the yz plane. (b) The mean energy flow in a wave is the energy density in the box length c, cross-sectional area $1\ \mathrm{m}^2$.

mean energy in a box length c and cross-section $1\,\mathrm{m}^2$, or

$$\langle \mathscr{S} \rangle = (\varepsilon_0 \langle E^2 \rangle)c$$

as before.

In free space $\mathscr{S}$ is given both by $(\boldsymbol{E} \times \boldsymbol{B}/\mu_0)$ and $(\boldsymbol{E} \times \boldsymbol{H})$. It is not clear from this as to whether $\mathscr{S}$ in a magnetizable medium would be $(\boldsymbol{E} \times \boldsymbol{B}/\mu_0)$, where $\boldsymbol{B}/\mu_0 = \boldsymbol{H} + \boldsymbol{M}$, or $(\boldsymbol{E} \times \boldsymbol{H})$. We shall discuss this when we consider propagation in matter.

When considering the flow of electromagnetic energy between two media (Chapter 11) a useful parameter is found to be that of *wave impedance*, Z. This is defined for free space by

$$Z_0 = \frac{E_{0x}}{H_{0y}} = \frac{\mu E_{0x}}{B_{0y}} \tag{9.36}$$

which with (9.31) and (9.32) gives

$$Z_0 = \mu_0 c \tag{9.37}$$

Since E is measured in $\mathrm{V\,m^{-1}}$ and H in $\mathrm{A\,m^{-1}}$, the wave impedance has the same dimensions as a circuit impedance, and from (9.37)

$$Z_0 = 377\,\mathrm{ohm} \tag{9.38}$$

10

Electromagnetic waves in dielectrics

Consideration of the propagation of electromagnetic waves in matter can be conveniently divided into two cases: (1) dielectrics, i.e. media in which there are no conduction electrons, and (2) conductors. Many gases, liquids and solids are good dielectrics and we begin with the effects of an oscillating electric field on such dielectrics. We then solve Maxwell's equations for linear, isotropic dielectrics and finally discuss the absorption and dispersion of electromagnetic waves in polar and non-polar dielectrics.

10.1 POLARIZATION OF DIELECTRICS

In electrostatics an applied electric field $\boldsymbol{E}$ produces polarization charges in dielectric media and a polarization $\boldsymbol{P}$ (7.12) related to the electric field by

$$\boldsymbol{P} = \varepsilon_0 \chi_e \boldsymbol{E} \tag{10.1}$$

where $\chi_e = (\varepsilon_r - 1)$ is the electric susceptibility of the dielectric and ε_0 is inserted to make χ_e dimensionless. The polarization of a dielectric is defined in terms of its internal dipole moments $\boldsymbol{p}$ by

$$\boldsymbol{P} = N\boldsymbol{p} \tag{10.2}$$

when there are N dipoles per unit volume. For a non-polar gas these dipole moments will be induced by the applied electric field and the *polarizability* α of the gas is given by

$$\boldsymbol{P} = \alpha \boldsymbol{E} \tag{10.3}$$

A simple atomic model for such a gas is that of electrons bound to their nuclei such that any displacement by an applied electric field is balanced by a restoring force proportional to that displacement.

For this model, application of a linearly polarized electromagnetic wave with electric vector of magnitude

$$E_x = E_0 \exp(\mathrm{i}\omega t) \tag{10.4}$$

would induce a damped, simple harmonic motion in the bound electrons given by

$$-eE_x = m(\ddot{x} + \gamma\dot{x} + \omega_0^2 x) \tag{10.5}$$

where x is the displacement parallel to E_x, γ is the damping constant of the electronic oscillators of natural frequency ω_0 and the effect of the magnetic component of the electromagnetic wave is assumed to be negligible. The steady-state displacements will be the dynamic responses to the incident wave, after initial transients,

$$x = x_0 \exp(\mathrm{i}\omega t)$$

and so

$$\dot{x} = \mathrm{i}\omega x, \quad \ddot{x} = -\omega^2 x$$

Hence

$$\boldsymbol{x} = \frac{-e}{m(\omega_0^2 - \omega^2 + \mathrm{i}\gamma\omega)}\boldsymbol{E} \tag{10.6}$$

The instantaneous dipole moment due to the displacement of the electron is

$$\boldsymbol{p} = -e\boldsymbol{x} \tag{10.7}$$

so that combining (10.6) and (10.7) and comparing with (10.3), we have the polarizability

$$\alpha(\omega) = \frac{e^2}{m(\omega_0^2 - \omega^2) + \mathrm{i}\gamma\omega)} \tag{10.8}$$

Clearly the polarizability is now a frequency-dependent parameter, and (10.2), (10.3) show that the polarization is

$$\boldsymbol{P} = N\alpha(\omega)\boldsymbol{E} \tag{10.9}$$

It is possible to calculate $\alpha(\omega)$ for particular gases, using perturbation theory, as shown in texts on quantum mechanics (such as *Quantum Mechanics* by P.C.W. Davies in this series).

These relations are true for dilute gases, where the atoms are so far apart as to be independent of one another. In dense gases, liquids and solids the polarizing field on each atom is not the external field $\boldsymbol{E}$, but a local field $\boldsymbol{E}_{loc}$, which allows for the polarization of neighbouring atoms. For dielectrics whose molecules do not have permanent dipole moments, that is non-polar dielectrics, a simple model is to assume that the local field is

$$\boldsymbol{E}_{loc} = \boldsymbol{E} + \boldsymbol{E}_{out} + \boldsymbol{E}_{in}$$

where $\boldsymbol{E}_{out}$ is the field due to the polarization of the dielectric outside a sphere of radius r_s, large compared with the intermolecular spacing a, surrounding the molecule, and $\boldsymbol{E}_{in}$ is the field due to the neighbouring molecules inside the sphere, as shown in Fig. 10.1(a). The polarized dielectric produces a surface charge density σ_p, which varies with the angle θ of a surface segment $dS = 2\pi \cdot r_s \sin\theta \cdot r_s \, d\theta$, such that

$$\sigma_p \, dS = \boldsymbol{P} \cdot \mathbf{d}\boldsymbol{S} = -P\cos\theta \, dS \tag{10.10}$$

where $\mathbf{d}\boldsymbol{S}$ is in the direction of the outward normal to the charged surface. The field $\boldsymbol{E}_{out}$ is parallel to $\boldsymbol{E}$ and is therefore along the x-axis in Fig. 10.1(b), so that

$$\boldsymbol{E}_{out} = -(\sigma_p \, dS)\cos\theta/(4\pi\varepsilon_0 r_s^2) \tag{10.11}$$

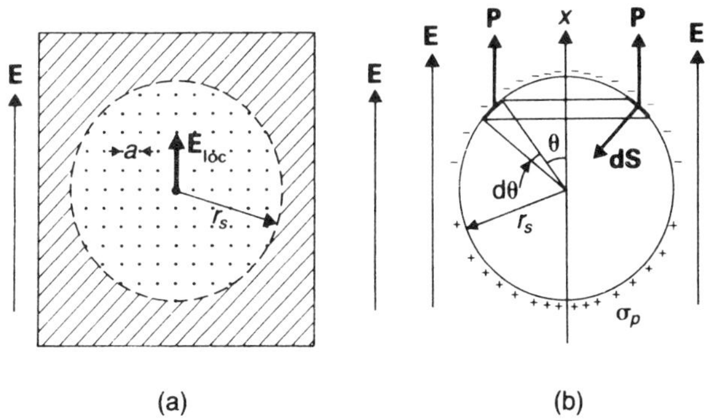

Fig. 10.1 (a) The local field $\boldsymbol{E}_{loc}$ in a dense dielectric on a molecule at O can be computed from the sum of the external field $\boldsymbol{E}$, the field due to the polarized dielectric (shaded) $\boldsymbol{E}_{out}$ and the field $\boldsymbol{E}_{in}$ due to the neighbouring molecules inside a sphere of radius $r_s \gg a$, the intermolecular distance. (b) The field $\boldsymbol{E}_{out}$ is equivalent to that of a hollow sphere with a surface charge density σ_p due to the polarized dielectric.

Combining (10.10) and (10.11) and integrating over the surface of the sphere

$$\boldsymbol{E}_{\text{out}} = \frac{\boldsymbol{P}}{2\varepsilon_0}\int_0^{\pi} \cos^2\theta \sin\theta \, \mathrm{d}\theta = \frac{\boldsymbol{P}}{3\varepsilon_0} \tag{10.12}$$

The field $\boldsymbol{E}_{\text{in}}$ due to the neighbouring molecules is more difficult to determine. Lorentz showed that for molecules on a simple cubic lattice $\boldsymbol{E}_{\text{in}} = 0$. It is reasonable to suppose that a random array of molecules, as in a glass or a liquid, should similarly have $\boldsymbol{E}_{\text{in}} = 0$. Therefore in this model

$$\boldsymbol{E}_{\text{loc}} = \boldsymbol{E} + \frac{\boldsymbol{P}}{3\varepsilon_0} \tag{10.13}$$

Similar arguments apply to the total electric field due to an electromagnetic wave, provided its wavelength $\lambda \gg a$, the intermolecular spacing. Therefore the dynamic polarization of a dense dielectric differs from that of a gas (10.9) and becomes

$$\boldsymbol{P} = N\alpha(\omega)\{\boldsymbol{E} + \boldsymbol{P}/(3\varepsilon_0)\} \tag{10.14}$$

Solving this equation for α and using (10.1) we obtain

$$\left(\frac{N\alpha}{1 - \dfrac{N\alpha}{3\varepsilon_0}}\right) = \frac{\boldsymbol{P}}{\boldsymbol{E}} = \varepsilon_0\chi_{\text{e}} = \varepsilon_0(\varepsilon_{\text{r}} - 1)$$

which can be rearranged to give the *Clausius–Mossotti equation*:

$$\frac{N\alpha}{3\varepsilon_0} = \left(\frac{\varepsilon_{\text{r}} - 1}{\varepsilon_{\text{r}} + 2}\right) \tag{10.15}$$

This equation approximates well to experiment for non-polar liquids and solids, showing that the local field model is valid over a wide range of densities. For example, for argon gas $\varepsilon_{\text{r}} = 1.000\,545$ at NTP, but liquid argon at 87 K is 780 times as dense and has $\varepsilon_{\text{r}} = 1.54$, compared with 1.50 that would be computed from (10.15).

In polar gases the effect of an applied electric field is to align the permanent dipole moments, $\boldsymbol{p}_0$, which otherwise are randomly orientated by the thermal motion of the molecules. The balance between the thermal agitation and the electrical alignment is similar to that for paramagnetics in a magnetic field and results in Curie's law for the polarization,

$$\boldsymbol{P}=\left(\frac{Np_0^2}{3k_\mathrm{B}T}\right)\boldsymbol{E}_\mathrm{loc} \tag{10.16}$$

where k_B is the Boltzmann constant and T the absolute temperature. Since the electronic polarization of the Clausius–Mossotti model still takes place, the total polarization will be the sum of (10.14) and (10.16). Therefore for polar gases the Clausius–Mossotti equation (10.15) becomes

$$\left(\frac{\varepsilon_\mathrm{r}-1}{\varepsilon_\mathrm{r}+2}\right)=\frac{N}{3\varepsilon_0}\left(\alpha+\frac{p_0^2}{3k_\mathrm{B}T}\right) \tag{10.17}$$

Measurements of the permittivity as the function of temperature thus enable both the polarizability α and the dipole moment p_0 of the molecules in a polar dielectric to be obtained. However, unlike paramagnetics, polar molecules are not rotated in solid dielectrics by an electric field, since the intermolecular forces between the electric dipoles in a solid are too strong to be overcome by external fields. Even in a polar liquid the Lorentz approximation for the local field does not apply and so (10.17) is limited to polar gases.

10.2 WAVE PARAMETERS IN DIELECTRICS

By dielectric media we mean gases, liquids or solids in which there are no free charges ($\rho_\mathrm{f}=j_\mathrm{f}=0$) and in which magnetization is negligible ($\boldsymbol{M}=0$). Maxwell's equations in dielectric media are then

$$\operatorname{div}\boldsymbol{E}=-\operatorname{div}\boldsymbol{P}/\varepsilon_0 \tag{10.18}$$

$$\operatorname{div}\boldsymbol{B}=0 \tag{10.19}$$

$$\operatorname{curl}\boldsymbol{E}=-\frac{\partial\boldsymbol{B}}{\partial t} \tag{10.20}$$

$$\operatorname{curl}\boldsymbol{B}=\frac{1}{c^2}\frac{\partial}{\partial t}\left(\frac{\boldsymbol{P}}{\varepsilon_0}+\boldsymbol{E}\right) \tag{10.21}$$

Here (10.18) follows from (7.1), (7.10) and (7.12), (10.19) and (10.20) are unchanged from (7.2) and (7.3), while (10.21) is (7.4) combined with (7.21) and $\mu_0\varepsilon_0=c^{-2}$. In this form the equations are quite general and can be used for anisotropic and nonlinear dielectrics. However, they are simplified for *isotropic* materials, in which there will be a uniform polarization and so $\operatorname{div}\boldsymbol{P}=-\rho_\mathrm{p}=0$, and for *linear* materials in which $\boldsymbol{P}$ will be proportional to $\boldsymbol{E}$ in amplitude, as well as being in the same direction as $\boldsymbol{E}$.

For such isotropic, linear dielectrics we can deduce a wave equation that is very similar to (9.3) for free space. As before, we have

$$\text{curl curl}\,\boldsymbol{E} = -\frac{\partial}{\partial t}(\text{curl}\,\boldsymbol{B})$$

and

$$\text{curl curl}\,\boldsymbol{E} = \text{grad div}\,\boldsymbol{E} - \nabla^2 \boldsymbol{E} = -\nabla^2 \boldsymbol{E}$$

Therefore

$$\nabla^2 \boldsymbol{E} - \frac{1}{c^2}\frac{\partial}{\partial t^2}\left(\frac{\boldsymbol{P}}{\varepsilon_0} + \boldsymbol{E}\right) = 0 \tag{10.22}$$

Since, by (7.24), $\boldsymbol{D} = \varepsilon_r \varepsilon_0 \boldsymbol{E}$ for linear, isotropic dielectrics and the electric displacement is defined by $\boldsymbol{D} = \varepsilon_0 \boldsymbol{E} + \boldsymbol{P}$ (2.30), a simple form of (10.22) is

$$\nabla^2 \boldsymbol{E} = \frac{\varepsilon_r}{c^2}\frac{\partial^2 E}{\partial t^2} \tag{10.23}$$

Following the solutions in section 9.2 of the similar wave equation in free space, we see that for a linearly-polarized plane wave

$$E_x = E_0 \exp \mathrm{i}(\omega t - kz) \tag{10.24}$$

where now the phase velocity

$$v = \frac{\omega}{k} = \frac{c}{\sqrt{\varepsilon_r}} \tag{10.25}$$

In physical optics we define the *refractive index*, n, of a medium as the ratio of the phase velocities of an electromagnetic wave in free space to that in the medium:

$$n = \frac{c}{v} = \sqrt{\varepsilon_r} \tag{10.26}$$

We have already seen that the permittivity of a dielectric is frequency dependent and so the refractive index will also vary with frequency. In particular we can rewrite the Clausius–Mossotti equation (10.15) as

$$\frac{n^2 - 1}{n^2 + 2} = \frac{N\alpha}{3\varepsilon_0} \tag{10.27}$$

and in this form it is often known as the *Lorentz–Lorenz equation* in studies of dielectric media at optical frequencies.

The magnetic vector of the electromagnetic wave in a dielectric bears a similar relation to the electric vector as that found for a plane wave in free space. Using Maxwell's third equation (10.20) it is easily seen that (9.18) becomes

$$\boldsymbol{B} = \frac{1}{v}(\hat{\mathbf{k}} \times \mathbf{E}) \tag{10.28}$$

where $\hat{\mathbf{k}}$ is a unit vector in the direction of propagation of the wave.

In dielectrics there can be losses, associated with the damping of the electronic oscillators in the nonpolar model (10.8) where $\alpha(\omega)$ is complex. For electromagnetic waves this is seen as an attenuation of the wave as it penetrates a dielectric. The combined effects of frequency dependence and absorption at optical frequencies are represented by a *complex refractive index*

$$n = n_{\text{R}} - \text{i}n_{\text{I}} \tag{10.29}$$

where the real part $n_{\text{R}} = c/v$ is the ordinary index of (10.26), while the imaginary part n_{I} corresponds to an attenuating wave. With this notation (10.24) for the wave becomes:

$$E_x = E_0 \exp \text{i}\omega\left(t - \frac{nz}{c}\right)$$

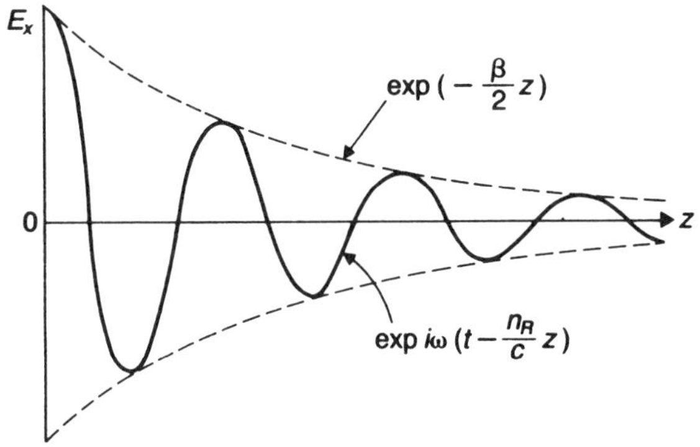

Fig. 10.2 The amplitude of an electromagnetic wave propagating along Oz at an instant of time has frequency $\omega/2\pi$, speed c/n_{R} and absorption coefficient $\beta/2$.

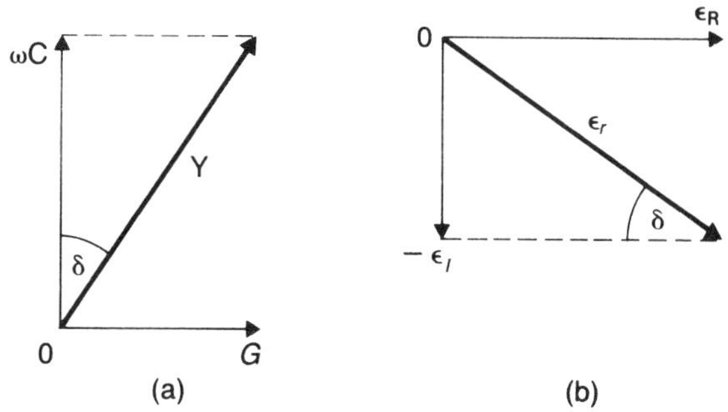

Fig. 10.3 A lossy dielectric has (a) an admittance $Y = G + i\omega C$, which is equivalent to (b) a complex permittivity where the loss tangent, $\tan\delta = G/\omega C = \varepsilon_I/\varepsilon_R$.

and

$$E_x = E_0 \exp i\omega\left(t - \frac{n_R z}{c}\right)\exp\left(-\frac{n_I \omega}{c} z\right) \tag{10.30}$$

The decaying amplitude of this wave is shown in Fig. 10.2, where β is the *absorption coefficient* derived from the intensity of the wave, proportional to E^2, decaying as $\exp(-\beta z)$, so that

$$\beta = 2n_I\omega/c \tag{10.31}$$

At radio frequencies the use of a *complex permittivity*,

$$\varepsilon_r = \varepsilon_R - i\varepsilon_I \tag{10.32}$$

is common and dielectric loss is often expressed by the *loss tangent*, $\tan\delta = \varepsilon_I/\varepsilon_R$, as seen on an Argand diagram (Fig. 10.3). Here a lossy dielectric is said to have an admittance $Y = G + i\omega C$, where G is its conductance and $B = \omega C$ its susceptance, so that $\tan\delta = G/\omega C$.

10.3 ABSORPTION AND DISPERSION

The behaviour of dielectrics over the electromagnetic spectrum varies enormously. A familiar contrast is the different result obtained for a non-polar gas like air and a polar liquid like water. For air we find the permittivity is approximately constant from measurements in a radio-frequency bridge to that deduced from the refractive index optically using the relation $n^2 = \varepsilon_r$ of (10.26), as shown in Table 10.1. On the other hand the permittivity of water at radio frequencies is far greater than that deduced from the refractive index. Most

Table 10.1 *Permittivities of common substances*

Frequency (Hz)	*10^6*	*5×10^{14}*
Air	$(\varepsilon_r - 1) = 567 \times 10^{-6}$	576×10^{-6}
Water	$\varepsilon_r = 80$	1.77

dielectrics exhibit resonances or relaxation peaks over the electromagnetic spectrum and we shall see how these can arise.

For nonpolar gases at low pressures, we start with the Lorentz–Lorenz equation (10.27), where α is given by (10.8), i.e.

$$\frac{n^2 - 1}{n^2 + 2} = \frac{Ne^2}{3m\varepsilon_0}\left(\frac{1}{\omega_0^2 - \omega^2 + \mathrm{i}\gamma\omega}\right) \tag{10.33}$$

The Lorentz correction (10.13) for a local field can be neglected in dilute gases and if we assume any resonance produces only a weak absorption line ($\gamma \ll \omega_0$), then we can simplify (10.33). Since by (10.26)

$$n^2 \simeq n_\mathrm{R}^2 = \varepsilon_r$$

and from (10.1), (10.2) and (10.3)

$$\varepsilon_r - 1 = \frac{\boldsymbol{P}}{\varepsilon_0 \boldsymbol{E}} = \frac{N\alpha}{\varepsilon_0}$$

(10.33) becomes

$$n_\mathrm{R}^2 = 1 + \frac{Ne^2}{m\varepsilon_0}\left\{\frac{\omega_0^2 - \omega^2}{(\omega_0^2 - \omega^2)^2 + \gamma^2\omega^2}\right\} \tag{10.34}$$

For a narrow absorption line (Fig. 10.4) the *natural width* is taken at the half-power points to be $2\Delta\omega$ and we can put

$$\omega_0^2 - \omega^2 = (\omega_0 + \omega)(\omega_0 - \omega) \simeq 2\omega(\omega_0 - \omega)$$

and rewrite (10.34) as

$$n_\mathrm{R}^2 = 1 + \frac{Ne^2}{2m\omega\varepsilon_0}\left\{\frac{(\omega_0 - \omega)}{(\omega_0 - \omega)^2 + (\Delta\omega)^2}\right\} \tag{10.35}$$

since $\gamma\omega = (\omega_0^2 - \omega^2) = 2\omega\Delta\omega$ at the half-power points. Similarly for n_I we have $n_\mathrm{I} \ll n_\mathrm{R}$, $n_\mathrm{R} \simeq 1$, so that

$$\varepsilon_\mathrm{I} = 2n_\mathrm{R}n_\mathrm{I} \simeq 2n_\mathrm{I}$$

and therefore

$$n_I = \frac{Ne^2}{4m\omega\varepsilon_0}\left\{\frac{\Delta\omega}{(\omega_0-\omega)^2+(\Delta\omega)^2}\right\} \tag{10.36}$$

The real and imaginary parts of the complex refractive index near a narrow absorption line are shown in Fig. 10.4, where the real part exhibits the characteristic dispersion shape and the imaginary part shows a Lorentzian shape. In optics the line width is the total width ($2\Delta\omega$) at the half-power points, whereas microwave spectroscopists commonly speak of the 'half-width' of the line ($\Delta\omega$) at the half-power points.

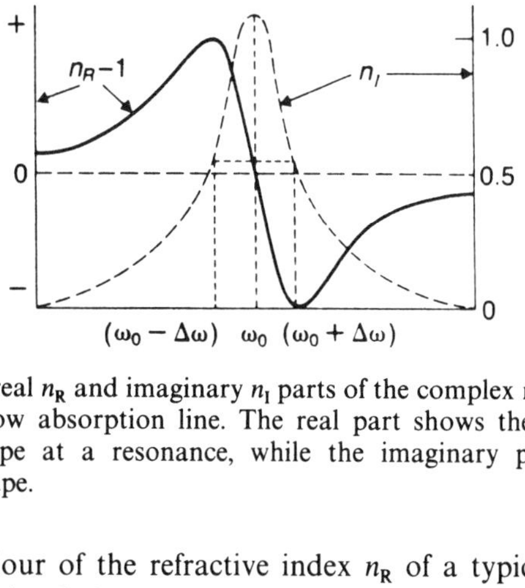

Fig. 10.4 The real n_R and imaginary n_I parts of the complex refractive index n near a narrow absorption line. The real part shows the characteristic dispersion shape at a resonance, while the imaginary part exhibits a Lorentzian shape.

The behaviour of the refractive index n_R of a typical molecular gas over much of the electromagnetic spectrum is illustrated in Fig. 10.5. At very low frequencies, or long wavelengths, we are measuring n_R in the range $\omega \ll \omega_0$, where ω_0 is any resonant frequency of absorption in the molecule. This is the region where n_R is a maximum. At shorter wavelengths it passes through a succession of resonances, each of which has an absorption peak in n_I (Fig. 10.4). When the range of wavelengths does not include an absorption peak, n_R increases as λ decreases and this is known as *normal dispersion*. On the other hand, in the absorption regions n_R decreases as λ decreases, and such regions are said to exhibit *anomalous dispersion*. At the longer wavelengths the absorption peaks are associated with internal motions of the atoms in the molecules, such as rotations and vibrations, while at the shorter wavelengths

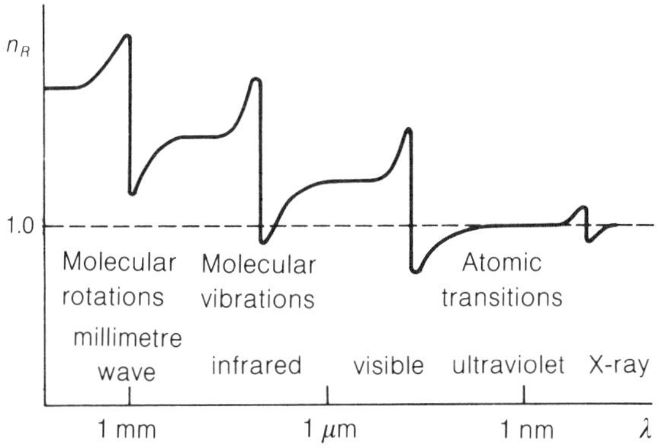

Fig. 10.5 The refractive index of a molecular gas exhibits a variety of resonances over a wide spectrum and these are associated with molecular and atomic transitions. The resonances shown are not to scale, but illustrate the spectral regions for each type of resonance.

the absorption is due to electronic transitions within the atoms. A full explanation of these absorption processes requires quantum mechanics, but a knowledge of elementary atomic physics is sufficient to understand that electronic transitions at optical wavelengths will be those of the outer shell, or valence, electrons, while the innermost shell, or core, electrons will produce resonant absorption at X-ray wavelengths.

At the high-frequency limit, $\omega \gg \omega_0$, all the Z electrons in an atom can be regarded as free, so that (10.32) becomes

$$n_R^2 = 1 - \frac{ZNe^2}{m\varepsilon_0\omega^2}$$

and, since $n_R \simeq 1$,

$$n_R = 1 - (ZNe^2/2m\varepsilon_0\omega^2) \tag{10.37}$$

Thus the refractive index is very close to, but slightly less than, unity for X-rays and γ-rays, and this is still true for dense media, such as metals, as we shall see in Chapter 12.

A different absorption process occurs in polar liquids, such as water, which show dielectric relaxation. When an electrostatic field is applied to a polar liquid the Brownian motion of the molecules acts to prevent the free rotation of the molecular dipoles and so dominates their motion. Removal of the electrostatic field then results in this polarization decaying with a *relaxation time*, τ, which is characteristic of the Brownian motion at a particular temperature, T. The Lorentz correction for the local field in polar gases (10.13) does not apply to polar liquids, since normal electric fields produce only a small polarization of the thermally agitated molecules. We therefore have, from (10.2), (10.3) and (10.16), that

$$\frac{\boldsymbol{P}}{\boldsymbol{E}} = N\left(\alpha + \frac{p_0^2}{3k_{\mathrm{B}}T}\right) \tag{10.38}$$

Here the first term represents the instantaneously induced electronic polarization ($\boldsymbol{P}_{\mathrm{i}}$) and the second is the time-dependent rotational polarization ($\boldsymbol{P}_{\mathrm{r}}$) of the molecular dipoles. Thus

$$\boldsymbol{P} = \boldsymbol{P}_{\mathrm{i}} + \boldsymbol{P}_{\mathrm{r}} \tag{10.39}$$

where the latter increases exponentially to a saturation value $\boldsymbol{P}_\infty$ that depends on the applied field:

$$\boldsymbol{P}_{\mathrm{r}} = \boldsymbol{P}_\infty[1 - \exp(-t/\tau)]$$

The rate of increase of the rotational polarization is

$$\frac{\mathrm{d}\boldsymbol{P}_{\mathrm{r}}}{\mathrm{d}t} = \frac{\boldsymbol{P}_\infty \exp(-t/\tau)}{\tau} = \frac{\boldsymbol{P}_\infty - \boldsymbol{P}_{\mathrm{r}}}{\tau} \tag{10.40}$$

or

$$\boldsymbol{P}_\infty = \boldsymbol{P}_{\mathrm{r}} + \frac{\tau\,\mathrm{d}\boldsymbol{P}_{\mathrm{r}}}{\mathrm{d}t} = \left(\frac{Np_0^2}{3k_{\mathrm{B}}T}\right)\boldsymbol{E} \tag{10.41}$$

When a radio-frequency field $E_0 \exp(\mathrm{i}\omega t)$ is applied the rotational polarization is

$$\boldsymbol{P}_{\mathrm{r}}(t) = \boldsymbol{P}_{\mathrm{r}}(0)\exp(\mathrm{i}\omega t)$$

so that (10.41) becomes

$$\boldsymbol{P}_{\mathrm{r}}(t)(1 + \mathrm{i}\omega\tau) = \left(\frac{Np_0^2}{3k_{\mathrm{B}}T}\right)\boldsymbol{E}_0 \exp(\mathrm{i}\omega t) \tag{10.42}$$

Over a range of radio frequencies the permittivity relaxes (Fig. 10.6) from its static value ε_{s}, to its high frequency value, ε_∞. From (10.39)

$$\boldsymbol{P} = \varepsilon_0(\varepsilon_{\mathrm{s}} - 1)\boldsymbol{E}$$

and

$$\boldsymbol{P}_{\mathrm{i}} = \varepsilon_0(\varepsilon_\infty - 1)\boldsymbol{E} \tag{10.43}$$

so that

$$\boldsymbol{P}_\infty = \varepsilon_0(\varepsilon_{\mathrm{s}} - \varepsilon_\infty)\boldsymbol{E} = \left(\frac{Np_0^2}{3k_{\mathrm{B}}T}\right)\boldsymbol{E} \tag{10.44}$$

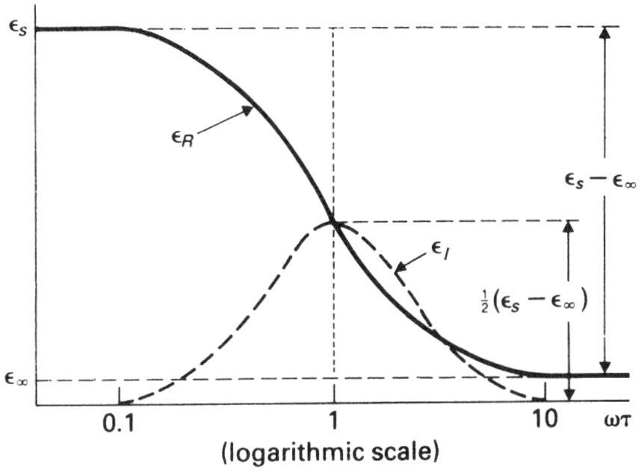

Fig. 10.6 The real ε_R and imaginary ε_I parts of the complex permittivity ε showing a broad Debye relaxation over a wide frequency range.

In the presence of a high-frequency field the permittivity is complex and

$$\boldsymbol{P}(t) = \varepsilon_0(\varepsilon_r - 1)\,\boldsymbol{E}(t) \tag{10.45}$$

Combining the last four equations, we find

$$\varepsilon_r = \varepsilon_\infty + \frac{\varepsilon_s - \varepsilon_\infty}{1 + i\omega\tau}$$

or, using (10.32),

$$\varepsilon_R = \varepsilon_\infty + \left(\frac{\varepsilon_s - \varepsilon_\infty}{1 + \omega^2\tau^2}\right) \tag{10.46}$$

$$\varepsilon_I = \frac{(\varepsilon_s - \varepsilon_\infty)\omega\tau}{(1 + \omega^2\tau^2)} \tag{10.47}$$

These are the *Debye equations* and are plotted in Fig. 10.6 with the parameter $\omega\tau$ on a logarithmic scale. Measurements over a range of microwave frequencies were necessary for water, where $\varepsilon_s = 80$ and $\tau = 10$ ps. However, the relaxation peak did not bring the value of ε_∞ down to n_R^2 measured optically, showing that absorption peaks of the type plotted in Fig. 10.5 occur in the infrared spectrum of water. It should be noted that a relaxation peak is very broad in frequency and that its height (ε_I) is less than the fall in the permittivity, ε_R.

11

Reflection and refraction

When electromagnetic waves are incident on the interface between two dielectrics the familiar phenomena of reflection and refraction take place. In this chapter we show that the wave properties of electromagnetic waves lead to the laws of reflection and refraction at plane surfaces, while their electromagnetic properties with the boundary conditions for electric and magnetic fields at dielectric interfaces lead to Fresnel's equations for plane-polarized waves. We conclude with a discussion of the special properties associated with waves incident at the Brewster angle and at angles greater than the critical angle.

11.1 BOUNDARY RELATIONS

In Chapters 2 and 6 Gauss's laws for the fluxes of $\boldsymbol{D}$ and $\boldsymbol{B}$ and the circulation laws for $\boldsymbol{E}$ and $\boldsymbol{H}$ were used to show that at a boundary between two media of permittivities $\varepsilon_1, \varepsilon_2$ and permeabilities μ_1, μ_2 the normal components of $\boldsymbol{D}$ and $\boldsymbol{B}$ together with the tangential components of $\boldsymbol{E}$ and $\boldsymbol{H}$, are continuous. Following Feynman these boundary relations will be deduced by applying Maxwell's equations to a plane interface where there is a sharp discontinuity in material properties, that is ε and μ change within a fraction of a wavelength of an electromagnetic wave.

For a polarizable, magnetizable medium with *no free charges* we have, from the alternative form of Maxwell's equations (7.13), (7.2), (7.3) and (7.22):

$$\operatorname{div} \boldsymbol{D} = 0 \tag{11.1}$$

$$\operatorname{div} \boldsymbol{B} = 0 \tag{11.2}$$

$$\text{curl}\,\boldsymbol{E} = -\,\partial\boldsymbol{B}/\partial t \tag{11.3}$$

$$\text{curl}\,\boldsymbol{H} = \partial\boldsymbol{D}/\partial t \tag{11.4}$$

We assume that the plane interface between the two media is the xy plane, so that (Fig. 11.1) the z-axis is normal to the interface. The physical properties at the boundary change over a distance $d \ll \lambda$, the wavelength of the electromagnetic wave that is to travel from medium 1 into medium 2. In Fig. 11.1(a) we have shown the electric displacement changing from D_1 to D_2 across the interface, but similar relations could exist for B, E and H.

At the boundary the three spatial coefficients of the components of these electromagnetic vectors behave very differently. The $\partial/\partial x$ and $\partial/\partial y$ coefficients of the field components will not change abruptly, while the $\partial/\partial z$ coefficients could peak sharply (Fig. 11.1(b)) if one of the field components changed rapidly during the narrow interface of width 'd'. So in applying Maxwell's equations we will consider the $\partial/\partial z$ coefficients only, as they dominate the interface.

From (11.1), we have

$$\frac{\partial D_x}{\partial x} + \frac{\partial D_y}{\partial y} + \frac{\partial D_z}{\partial z} = 0$$

so that, since $\partial D_x/\partial x = \partial D_y/\partial y = 0$, then $\partial D_z/\partial z$ must be zero at the interface and there can be no peak like Fig. 11.1(b). Therefore

$$D_{1z} = D_{2z} \tag{11.5}$$

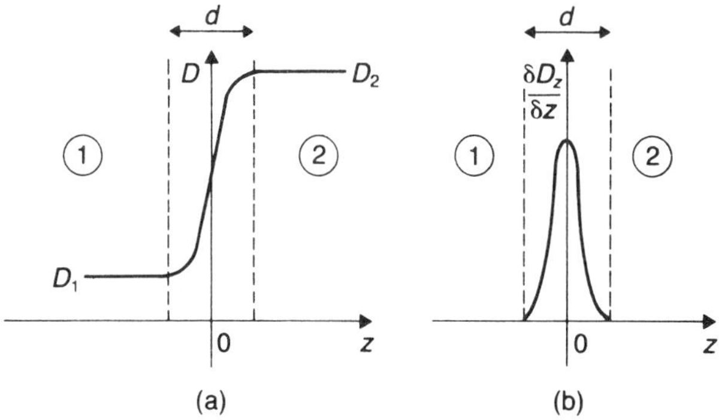

Fig. 11.1 (a) The electric displacement $\boldsymbol{D}$ at the interface of width d between two dielectrics 1, 2 is postulated to change from D_1 to D_2. (b) In consequence the z-component to div $\boldsymbol{D}$ peaks at the boundary.

Similarly, from (11.2)

$$B_{1z} = B_{2z} \tag{11.6}$$

On the other hand, (11.3) and (11.4) are vector equations where each vector component must be equal, giving, for (11.3),

$$\frac{\partial E_z}{\partial y} - \frac{\partial E_y}{\partial z} = -\frac{\partial B_x}{\partial t}$$

$$\frac{\partial E_x}{\partial z} - \frac{\partial E_z}{\partial x} = -\frac{\partial B_y}{\partial t}$$

$$\frac{\partial E_y}{\partial x} - \frac{\partial E_x}{\partial y} = -\frac{\partial B_z}{\partial t}$$

In these equations only the components $\partial E_y/\partial z$ and $\partial E_x/\partial z$ could peak sharply at the interface, but the time derivatives of B will not have sharp peaks. Therefore E_y and E_x must be continuous at the interface and

$$E_{1y} = E_{2y} \tag{11.7}$$

$$E_{1x} = E_{2x} \tag{11.8}$$

Similarly, from (11.4)

$$H_{1y} = H_{2y} \tag{11.9}$$

$$H_{1x} = H_{2x} \tag{11.10}$$

Equations (11.5) to (11.10) show which components of $\boldsymbol{D}$, $\boldsymbol{B}$, $\boldsymbol{E}$ and $\boldsymbol{H}$ are continuous across an interface and correspond exactly with the equations in Chapters 2 and 6, where the normal components of $\boldsymbol{D}$ and $\boldsymbol{B}$ and the tangential components of $\boldsymbol{E}$ and $\boldsymbol{H}$ were continuous at a boundary. We shall apply these boundary conditions to the electromagnetic fields of a plane wave to determine the Fresnel equations, but first we consider the relationship of the wave properties of an electromagnetic wave across an interface (Fig. 11.2).

It is easily shown from Maxwell's equations, (11.1) to (11.4), that the wave equations for an electromagnetic wave in a polarizable, magnetizable medium are (Chapter 10, exercise 2)

$$\nabla^2 \boldsymbol{E} = \frac{1}{v^2}\frac{\partial^2 \boldsymbol{E}}{\partial t^2} \tag{11.11}$$

$$\nabla^2 \boldsymbol{B} = \frac{1}{v^2}\frac{\partial^2 \boldsymbol{B}}{\partial t^2} \tag{11.12}$$

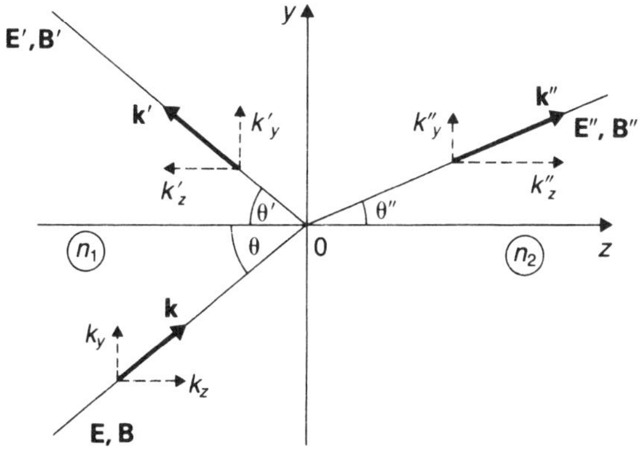

Fig. 11.2 An electromagnetic wave incident at an angle θ in the yz plane to an interface in the xy plane separating two dielectric media of refractive indices n_1 and n_2 produces a reflected wave at angle θ' and a transmitted wave at angle θ''.

where $v^2 = c^2/(\mu_r \varepsilon_r) = c^2/n^2$. For a plane wave of wave vector $\boldsymbol{k}$ and frequency ω in such a medium, we have, by comparison with (10.24) and (10.28), therefore

$$\begin{aligned} \boldsymbol{E} &= \boldsymbol{E}_0 \exp \mathrm{i}(\omega t - \boldsymbol{k}\cdot\boldsymbol{r}) \\ \boldsymbol{B} &= \frac{\hat{\boldsymbol{k}} \times \boldsymbol{E}}{v} = \frac{\boldsymbol{k} \times \boldsymbol{E}}{\omega} \end{aligned} \tag{11.13}$$

where $\boldsymbol{E}, \boldsymbol{B}$ are the electric and magnetic vectors of the plane wave at a point $\boldsymbol{r}$ from the origin at time t. Following its interaction with the xy plane of the surface (Fig. 11.2) the reflected wave is

$$\begin{aligned} \boldsymbol{E}' &= \boldsymbol{E}_0' \exp \mathrm{i}(\omega' t - \boldsymbol{k}'\cdot\boldsymbol{r}) \\ \boldsymbol{B}' &= (\boldsymbol{k}' \times \boldsymbol{E}')/\omega' \end{aligned} \tag{11.14}$$

and the transmitted wave is

$$\begin{aligned} \boldsymbol{E}'' &= \boldsymbol{E}_0'' \exp \mathrm{i}(\omega'' t - \boldsymbol{k}''\cdot\boldsymbol{r}) \\ \boldsymbol{B}'' &= (\boldsymbol{k}'' \times \boldsymbol{E}'')/\omega'' \end{aligned} \tag{11.15}$$

If we choose axes such that the incident wave vector $\boldsymbol{k}$ is in the yz plane, then

$$\boldsymbol{k}\cdot\boldsymbol{r} = k_y y + k_z z \tag{11.16}$$

At the interface ($z = 0$) the sum of the incident and reflected electric fields must equal that transmitted, so that

$$\dot{\boldsymbol{E}}_0 \exp \mathrm{i}(\omega t - k_y y) + \boldsymbol{E}'_0 \exp \mathrm{i}(\omega' t - k'_y y) = \boldsymbol{E}''_0 \exp \mathrm{i}(\omega'' t - k''_y y) \tag{11.17}$$

For this to be true at all times t and for all points $(y, 0)$ on the interface, clearly

$$\omega = \omega' = \omega'' \tag{11.18}$$

so that there can be no change of frequency occurring. Since the speed v_1 of the incident and reflected waves must be the same

$$\boldsymbol{k} \cdot \boldsymbol{k} = k^2 = \frac{\omega^2}{v_1^2} = k'^2 \tag{11.19}$$

and

$$k''^2 = \frac{v_1^2 k^2}{v_2^2} = \frac{n_2^2 k^2}{n_1^2} \tag{11.20}$$

But for (11.17) to be true for all y

$$k_y = k'_y = k''_y \tag{11.21}$$

To satisfy both (11.19) and (11.21) the reflected wave must have $k'_z = -k_z$ and so (Fig. 11.2) the angle of incidence θ equals the angle of reflection θ' and is in the yz plane. Electromagnetic waves therefore obey the *laws of reflection*.

For the transmitted wave, (11.20) and (11.21) give

$$k''^2_z = \left(\frac{n_2^2}{n_1^2}\right) k^2 - k_y^2 \tag{11.22}$$

which is true for all dielectrics, including the conditions under which n is complex (10.29). When n_1 and n_2 are real, that is away from resonances (Fig. 11.5), then

$$k_y = k \sin\theta = k''_y = k'' \sin\theta''$$

or

$$\frac{\sin\theta''}{\sin\theta} = \frac{k}{k''} = \frac{n_1}{n_2} \tag{11.23}$$

which is the *law of refraction* discovered by Snell experimentally in 1621.

11.2 FRESNEL'S EQUATIONS

Although we have proved that electromagnetic waves obey the experimental laws of reflection and refraction at plane dielectric interfaces, these laws follow from any wave theory having the general wave equation (9.5). The distinctive features of electromagnetism are found in the amplitudes of the reflected and transmitted waves that fulfil the electromagnetic boundary conditions, (11.5) to (11.10). In general, as we saw in section 9.2, a plane electromagnetic wave is elliptically polarized, but any electric polarization can always be represented as the sum of an electric vector normal to the plane of incidence (Fig. 11.3(a)) and one parallel to it (Fig. 11.3(b)). The magnetic vectors then follow from the relations $\boldsymbol{B} = (\boldsymbol{k} \times \boldsymbol{E})/\omega$ for each wave.

11.2.1 *E* normal or TE polarization

In Fig. 11.3(a) the incident electric vector is normal to the yz plane in the $-x$ direction (out of the figure), so that the incident magnetic vector is in the yz plane (compare Fig. 9.3). For isotropic dielectrics the induced oscillations will be parallel to the incident ones, so that E', B' and E'', B'' are as shown. At the interface ($z = 0$), the superposition of the electric fields, (11.17), with (11.18) and (11.21) becomes

$$E_0 + E_0' = E_0'' \tag{11.24}$$

For the magnetic vectors in the yz plane only their y components, (11.9), provide any additional information. Since $\boldsymbol{B} = \mu_r \mu_0 \boldsymbol{H} =$

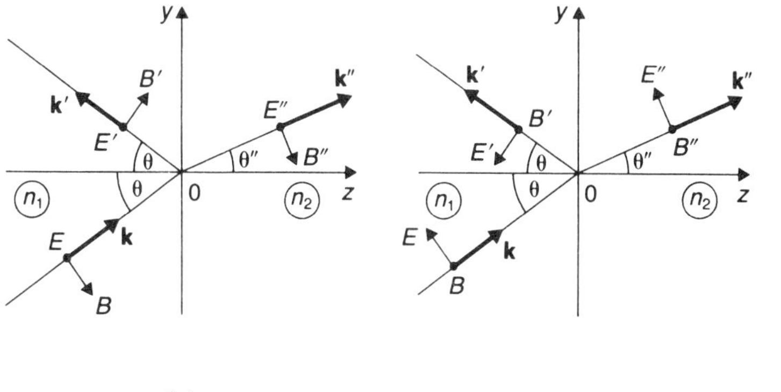

Fig. 11.3 (a) A plane polarized electromagnetic wave with the electric vectors *normal* to the plane of incidence and directed out of the figure is partially reflected and transmitted. (b) A plane-polarized electromagnetic wave with the electric vectors *parallel* to the plane of incidence is partially reflected and transmitted. The magnetic vectors are directed out of the figure.

$(\boldsymbol{k} \times \boldsymbol{E})/\omega$, (11.9) becomes

$$\frac{(\boldsymbol{k} \times \boldsymbol{E})_{1y}}{\mu_{r1}\mu_0\omega} = \frac{(\boldsymbol{k} \times \boldsymbol{E})_{2y}}{\mu_{r2}\mu_0\omega}$$

which simplifies to

$$\frac{(\hat{\mathbf{k}} \times \boldsymbol{E})_{1y}}{Z_1} = \frac{(\hat{\mathbf{k}} \times \boldsymbol{E})_{2y}}{Z_2} \tag{11.25}$$

since $k = \omega/v$ and where, from (9.36), the wave impedance is given by

$$Z = \frac{E_x}{H_y} = \mu_r\mu_0 v \tag{11.26}$$

Hence, in terms of the wave impedances of each medium

$$\frac{E_0 \cos\theta}{Z_1} - \frac{E'_0 \cos\theta}{Z_1} = \frac{E''_0 \cos\theta''}{Z_2} \tag{11.27}$$

Combining (11.24) and (11.27), the reflected and transmitted amplitudes for $\boldsymbol{E}$ normal to the plane of incidence or TE polarization are

$$\left(\frac{E'_0}{E_0}\right)_{\text{TE}} = \frac{Z_2 \cos\theta - Z_1 \cos\theta''}{Z_2 \cos\theta + Z_1 \cos\theta''} \tag{11.28}$$

and

$$\left(\frac{E''_0}{E_0}\right)_{\text{TE}} = \frac{2Z_2 \cos\theta}{Z_2 \cos\theta + Z_1 \cos\theta''} \tag{11.29}$$

At optical frequencies $\mu_{r1} = \mu_{r2} = 1$, $n^2 = \varepsilon_r$ and so $Z_1/Z_2 = v_1/v_2 = n_2/n_1 = \sin\theta/\sin\theta''$, by (11.23). Hence these amplitude equations simplify to

$$\left(\frac{E'_0}{E_0}\right)_{\text{TE}} = \frac{\sin(\theta'' - \theta)}{\sin(\theta'' + \theta)}, \quad \left(\frac{E''_0}{E_0}\right)_{\text{TE}} = \frac{2\sin\theta''\cos\theta}{\sin(\theta'' + \theta)} \tag{11.30}$$

11.2.2 *E* parallel or TM polarization

In Fig. 11.3(b) the incident magnetic vector is normal to the yz plane in the $-x$ direction, so that the incident electric vector is in the yz plane, as given by $\boldsymbol{B} = (\boldsymbol{k} \times \boldsymbol{E})/\omega$. The relevant boundary equations are now (11.7) and (11.10), giving

$$(E_0 - E'_0)\cos\theta = E''_0 \cos\theta'' \tag{11.31}$$

and

$$\frac{E_0}{Z_1} + \frac{E'_0}{Z_1} = \frac{E''_0}{Z_2} \tag{11.32}$$

Solving (11.31) and (11.32) for the reflected and transmitted amplitudes for ***E*** parallel to plane of incidence or TM polarization gives

$$\left(\frac{E'_0}{E_0}\right)_{\text{TM}} = \frac{-Z_2 \cos\theta'' + Z_1 \cos\theta}{Z_2 \cos\theta'' + Z_1 \cos\theta} \tag{11.33}$$

and

$$\left(\frac{E''_0}{E_0}\right)_{\text{TM}} = \frac{2Z_2 \cos\theta}{Z_2 \cos\theta'' + Z_1 \cos\theta} \tag{11.34}$$

At optical frequencies these simplify to:

$$\left(\frac{E'_0}{E_0}\right)_{\text{TM}} = \frac{\tan(\theta - \theta'')}{\tan(\theta + \theta'')}$$

$$\left(\frac{E''_0}{E_0}\right)_{\text{TM}} = \frac{2\cos\theta \sin\theta''}{\sin(\theta + \theta'')\cos(\theta - \theta'')}. \tag{11.35}$$

Equations (11.30) and (11.35), known as Fresnel's equations after their discoverer, apply at optical frequencies to transparent media, where the refractive indices are real. They are drawn for air to glass in Fig. 11.4.

11.2.3 Polarization by reflection

For a particular angle of incidence, known as the Brewster angle, the reflected wave for TM polarization has zero amplitude: the reflection disappears. The Brewster angle is, from (11.35), given by

$$(\theta_B + \theta'') = \pi/2$$

which, with Snell's law, becomes

$$n_1 \sin\theta_B = n_2 \sin\left(\frac{\pi}{2} - \theta_B\right) = n_2 \cos\theta_B$$

or

$$\tan\theta_B = \frac{n_2}{n_1} \tag{11.36}$$

At this angle an unpolarized wave would be reflected as a plane polarized wave with TE polarization, the TM polarization being

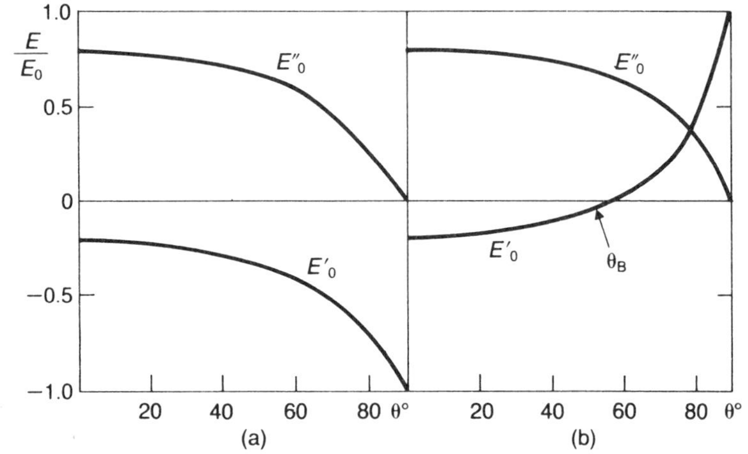

Fig. 11.4 The reflected (E'_0) and transmitted (E''_0) relative amplitudes from Fresnel's equations: (a) transverse electric (TE) polarization; (b) transverse magnetic (TM) polarization, for an air/glass interface with $n_2/n_1 = 1.5$. θ_B = Brewster angle.

fully transmitted. For air to glass, $n_2/n_1 = 1.5$ and the Brewster angle is 56°.

The phase of the reflected wave for TE polarization (Fig. 11.4(a)) depends only on n_2/n_1 and is always negative for $n_2 > n_1$. On the other hand for TM polarization the phase changes from negative to positive at the Brewster angle (Fig. 11.4(b)) so that its sign depends on the angle of incidence.

11.3 ENERGY FLOW AT A BOUNDARY

In section 9.4 we showed that the equation of conservation of energy flow was (9.25)

$$-\frac{\partial u}{\partial t} = \operatorname{div} \mathscr{S} + \boldsymbol{E} \cdot \boldsymbol{j}$$

where $\boldsymbol{j}$ was the current produced by a charge q moving at speed v, u was the energy density and $\mathscr{S}$ the flux of electromagnetic energy. In applying it to free space, we found that $\mathscr{S} = \boldsymbol{E} \times \boldsymbol{H}$, the Poynting vector. We will now show that it remains $\boldsymbol{E} \times \boldsymbol{H}$ in a medium.

The current $\boldsymbol{j}$ in a medium will be that due to the free charges j_f, given by (7.22), so that

$$\boldsymbol{E}\cdot\boldsymbol{j}=\boldsymbol{E}\cdot\operatorname{curl}\boldsymbol{H}-\boldsymbol{E}\cdot\frac{\partial\boldsymbol{D}}{\partial t} \tag{11.37}$$

Using the vector identity for $\operatorname{div}(\boldsymbol{A}\times\boldsymbol{B})$ from Appendix E

$$\boldsymbol{E}\cdot\operatorname{curl}\boldsymbol{H}=\operatorname{div}(\boldsymbol{H}\times\boldsymbol{E})+\boldsymbol{H}\cdot\operatorname{curl}\boldsymbol{E}$$

which, with Maxwell's equation for curl $\boldsymbol{E}$, becomes

$$\boldsymbol{E}\cdot\operatorname{curl}\boldsymbol{H}=\operatorname{div}(\boldsymbol{H}\times\boldsymbol{E})-\frac{\partial}{\partial t}\left(\frac{\boldsymbol{H}\cdot\boldsymbol{B}}{2}\right)$$

Substituting this expression in (11.37) gives

$$\boldsymbol{E}\cdot\boldsymbol{j}=\operatorname{div}(\boldsymbol{H}\times\boldsymbol{E})-\frac{\partial}{\partial t}\left\{\frac{\boldsymbol{E}\cdot\boldsymbol{D}}{2}+\frac{\boldsymbol{H}\cdot\boldsymbol{B}}{2}\right\} \tag{11.38}$$

Comparing (11.38) and (9.25), we see that for energy flow and energy density in a medium

$$\mathscr{S}=\boldsymbol{E}\times\boldsymbol{H} \tag{11.39}$$

$$u=\tfrac{1}{2}\{\boldsymbol{E}\cdot\boldsymbol{D}+\boldsymbol{H}\cdot\boldsymbol{B}\} \tag{11.40}$$

where for linear, isotropic media, $\boldsymbol{B}=\mu_r\mu_0\boldsymbol{H}$ and $\boldsymbol{D}=\varepsilon_r\varepsilon_0\boldsymbol{E}$. Although the Poynting vector has not changed, the energy density now includes the energy associated with the polarization current density $\partial\boldsymbol{P}/\partial t$ and the magnetic current density curl $\boldsymbol{M}$ (7.21) which were shown in Chapter 7 to lead to the Maxwell equation (7.21). That is why (11.40) for energy density in a medium differs from (9.30) for energy density in free space.

At optical frequencies in dielectrics we have n real, so that the $\boldsymbol{E}$ and $\boldsymbol{H}$ vectors in an electromagnetic wave are in phase and the electric and magnetic energy densities are equal

$$\tfrac{1}{2}\varepsilon_r\varepsilon_0E^2=\tfrac{1}{2}\mu_r\mu_0H^2 \tag{11.41}$$

The total energy density is therefore $\varepsilon_r\varepsilon_0E^2=n^2\varepsilon_0E^2$ and the average energy density $\langle u\rangle=n^2\varepsilon_0E^2_{\mathrm{rms}}$. Therefore the average Poynting vector

$$\langle\mathscr{S}\rangle=\tfrac{1}{2}E_0H_0\hat{\mathbf{k}} \tag{11.42}$$

so that the intensity

$$\langle\mathscr{S}\rangle=\frac{1}{2}\left(\frac{\varepsilon_r\varepsilon_0}{\mu_r\mu_0}\right)^{1/2}E_0^2=\frac{E_0^2}{2Z} \tag{11.43}$$

or

$$\langle \mathscr{S} \rangle = \tfrac{1}{2} v(\varepsilon_r \varepsilon_0) E_0^2 = v \langle u \rangle \tag{11.44}$$

as would be expected for the average energy flow across unit area.

The *reflection coefficient* or *reflectance* R is the ratio of the average energy flux per second reflected to that incident on an interface, so that, from (11.42) and (11.43)

$$R = \frac{\langle \mathscr{S}' \rangle \cdot \hat{\mathbf{n}}}{\langle \mathscr{S} \rangle \cdot \hat{\mathbf{n}}} = \frac{E_0'^2}{E_0^2} \tag{11.45}$$

where $\hat{\mathbf{n}}$ is a unit vector normal to the interface. Similarly the *transmission coefficient* or *transmittance* T is

$$T = \frac{\langle \mathscr{S}'' \rangle \cdot \hat{\mathbf{n}}}{\langle \mathscr{S} \rangle \cdot \hat{\mathbf{n}}} = \frac{Z_1 E_0''^2 \cos\theta''}{Z_2 E_0^2 \cos\theta} \tag{11.46}$$

When energy is conserved at the interface, we always have

$$R + T = 1 \tag{11.47}$$

Using Fresnel's equations for E_0', E_0'', we find for TE polarization

$$R_{\mathrm{TE}} = \left(\frac{Z_2 \cos\theta - Z_1 \cos\theta''}{Z_2 \cos\theta + Z_1 \cos\theta''} \right)^2 = \frac{\sin^2(\theta'' - \theta)}{\sin^2(\theta'' + \theta)} \tag{11.48}$$

and for TM polarization

$$R_{\mathrm{TM}} = \left(\frac{Z_2 \cos\theta'' - Z_1 \cos\theta}{Z_2 \cos\theta'' + Z_1 \cos\theta} \right)^2 = \frac{\tan^2(\theta - \theta'')}{\tan^2(\theta + \theta'')} \tag{11.49}$$

where the second expressions apply at optical frequencies where $Z_1/Z_2 = n_2/n_1 = \sin\theta/\sin\theta''$. In both cases the transmittances can be found using (11.47) or from (11.46), (11.29) and (11.34). The reflectances for air/glass with $n_2/n_1 = 1.5$ are shown in Fig. 11.5. At the Brewster angle $R_{\mathrm{TM}} = 0$, as expected.

At normal incidence ($\theta = 0$), both (11.48) and (11.49) give

$$R_0 = \left(\frac{Z_2 - Z_1}{Z_2 + Z_1} \right)^2 = \left(\frac{n_1 - n_2}{n_1 + n_2} \right)^2 \tag{11.50}$$

so that

$$T_0 = 1 - R_0 = \frac{4 Z_1 Z_2}{(Z_2 + Z_1)^2} = \frac{4 n_1 n_2}{(n_1 + n_2)^2} \tag{11.51}$$

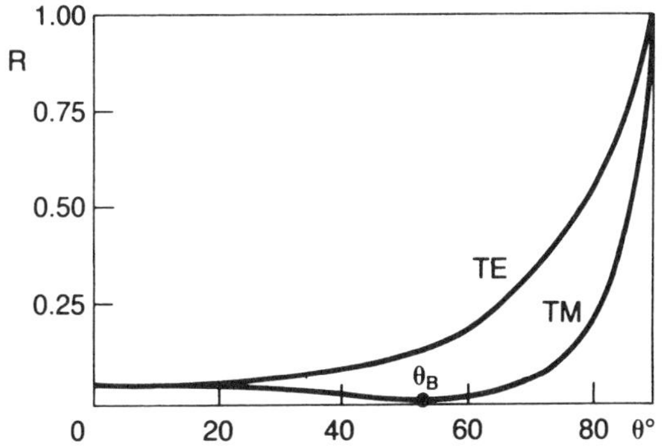

Fig. 11.5 The reflectances for an air/glass interface when transverse electric (TE) and transverse magnetic (TM) radiation is incident at angle θ. θ_B = Brewster angle.

11.4 TOTAL INTERNAL REFLECTION

When radiation is being transmitted from a dense to a less dense medium, for example glass to air, with $n_1 > n_2$, then Snell's law (11.23) gives

$$\sin\theta'' = \frac{n_1}{n_2}\sin\theta$$

There will therefore be a critical angle of incidence θ_c for which θ'' has its maximum value of 90°

$$\sin\theta_c = \frac{n_2}{n_1} \tag{11.52}$$

For incidence at all angles $\theta > \theta_c$, there can only be 'total internal reflection'.

This phenomenon can be understood if we go back to equation (11.22), which applies when n is complex

$$k_z''^2 = \left(\frac{n_2}{n_1}\right)^2 k^2 - k_y^2$$

Since $k^2 = \omega^2/v_1^2$, $k_y = k\sin\theta$, and $v_1^2/v_2^2 = n_2^2/n_1^2$

$$k_z''^2 = \frac{\omega^2}{v_2^2}\left(1 - \frac{n_1^2}{n_2^2}\sin^2\theta\right) \tag{11.53}$$

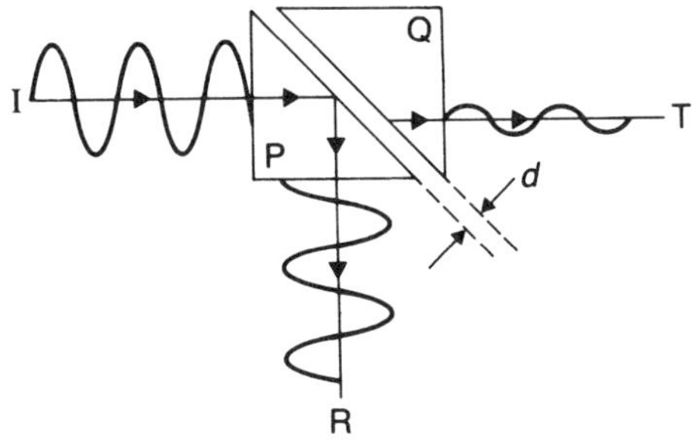

Fig. 11.6 An incident microwave beam I is internally reflected by the prism P to the receiver R, but a small transmitted signal can be observed at T when a second prism Q is at a distance $d < \lambda$, the microwave wavelength. This demonstrates that an evanescent wave accompanies total internal reflection.

Now when $\theta > \theta_c$, $n_1 \sin\theta/n_2 > 1$ and so k_z'' is an imaginary number, say $-ik_I$, showing that the amplitude is decaying in a similar way to imaginary parts of n and ε_r in (10.29) and (10.32). Hence the transmitted wave, (11.15), becomes

$$E'' = E_0'' \exp(-k_I z) \exp i(\omega t - k_y y) \tag{11.54}$$

and we see that it travels only a short distance into the second medium, decaying in amplitude by $1/e$ within $\lambda/2\pi$. The existence of this evanescent wave can be demonstrated most conveniently with microwaves (say $\lambda = 3\,\text{cm}$), as illustrated in Fig. 11.6, where the internally reflected signal at R will decrease when the second prism Q is brought within a distance $d < \lambda$ and a transmitted signal can be detected at T.

12

Electromagnetic waves in conductors

Electromagnetic waves propagate with their electric and magnetic fields oscillating about the direction of propagation (Fig. 9.6) so that when they interact with matter the largest effects come from the lightest charged particles, the electrons. In dielectrics the electrons are bound charges and in Chapter 10 we found that the polarization was characterized by the atomic polarizability $\alpha(\omega)$, given by the Clausius–Mossotti and Lorentz–Lorenz equations. In conductors these effects are still present, but they are normally very small compared with the interactions with the conduction electrons. The densities of these carriers (Table 12.1) varies from $9 \times 10^{28}\,\mathrm{m}^{-3}$ in a noble metal to about 10^{20}–$10^{24}\,\mathrm{m}^{-3}$ in semiconductors and dense plasmas, but can be as little as $10^{11}\,\mathrm{m}^{-3}$ in weakly ionized plasmas like the ionosphere, while interstellar gas has a density of about $10^{6}\,\mathrm{m}^{-3}$. Besides the density, the most important parameters for the propagation of electromagnetic waves are the relaxation time τ

Table 12.1 *Density of charge carriers in conductors*

Conductor	*Example*	*Density, N* (m^{-3})
Noble metal	Copper, silver, gold	6×10^{28}–9×10^{28}
Alkali metal	Sodium, caesium	8×10^{27}–5×10^{28}
Semi-metal	Bismuth, antimony, arsenic	3×10^{23}–2×10^{26}
Semiconductor	Extrinsic germanium	5×10^{20}–10^{24}
Dense plasma	Solar, laser, discharge	10^{20}–10^{26}
Weak plasma	Ionosphere, space	10^{6}–10^{11}

between collisions of the carriers, which determines the conductivity, and the frequency ($\omega/2\pi$) of the wave.

12.1 WAVE PARAMETERS IN CONDUCTORS

Classically, a conducting medium obeys Ohm's law,

$$\boldsymbol{j}_f = \gamma \boldsymbol{E} \tag{12.1}$$

where γ is the electrical conductivity (SI unit = S m^{-1}) of the medium, and has net charge density $\rho = 0$. Hence, Maxwell's equations in a conductor are

$$\operatorname{div} \boldsymbol{E} = 0 \tag{12.2}$$

$$\operatorname{div} \boldsymbol{B} = 0 \tag{12.3}$$

$$\operatorname{curl} \boldsymbol{E} = -\frac{\partial \boldsymbol{B}}{\partial t} \tag{12.4}$$

$$\operatorname{curl}\left(\frac{\boldsymbol{B}}{\mu_r \mu_0}\right) = \gamma \boldsymbol{E} + \varepsilon_r \varepsilon_0 \frac{\partial \boldsymbol{E}}{\partial t} \tag{12.5}$$

where the first three equations follow from (7.1), (7.2) and (7.3). The fourth equation, combining (7.22), (7.23), (7.24) and (12.1), is valid for linear, isotropic conductors. We follow and same procedure as in dielectrics, putting

$$\operatorname{curl}\operatorname{curl} \boldsymbol{E} = \operatorname{grad}\operatorname{div} \boldsymbol{E} - \nabla^2 \boldsymbol{E} = -\nabla^2 \boldsymbol{E}$$

and hence find

$$\nabla^2 \boldsymbol{E} = \mu_r \mu_0 \gamma \frac{\partial \boldsymbol{E}}{\partial t} + \mu_r \mu_0 \varepsilon_r \varepsilon_0 \frac{\partial^2 \boldsymbol{E}}{\partial t^2} \tag{12.6}$$

Similarly

$$\nabla^2 \boldsymbol{H} = \mu_r \mu_0 \gamma \frac{\partial \boldsymbol{H}}{\partial t} + \mu_r \mu_0 \varepsilon_r \varepsilon_0 \frac{\partial^2 \boldsymbol{H}}{\partial t^2} \tag{12.7}$$

In both equations, on the right-hand side the first term is derived from the conduction current and the second term from the displacement current.

As before, a linearly-polarized plane wave travelling along the z-axis could have as its electric vector

$$E_x = E_0 \exp \mathrm{i}(\omega t - kz) \tag{12.8}$$

where, from (12.6)

$$-k^2 = \mathrm{i}\,\omega\mu_r\mu_0\gamma - \omega^2\mu_r\mu_0\varepsilon_r\varepsilon_0$$

or

$$k^2 = \frac{\omega^2}{c^2}\left(1 - \frac{\mathrm{i}\gamma}{\omega\varepsilon_r\varepsilon_0}\right)(\mu_r\varepsilon_r) \tag{12.9}$$

The wave number is therefore complex and can be written:

$$k = k_R - \mathrm{i}k_I \tag{12.10}$$

so that the equation of the wave, (12.8), becomes

$$E_x = E_0 \exp\mathrm{i}\,(\omega t - k_R z)\exp(-k_I z) \tag{12.11}$$

This is an oscillatory field with decaying amplitude, similar to that shown in Fig. 10.2. However, in a *good conductor* $\gamma/\omega\varepsilon_r\varepsilon_0 \gg 1$ for frequencies up to at least the microwave range and so (12.9) and (12.10) yield (with $c^{-2} = \mu_0\varepsilon_0$) the simple result

$$k = \left(\frac{\mu_r\mu_0\omega\gamma}{2}\right)^{1/2}(1 - \mathrm{i}) \tag{12.12}$$

In this case the wave is heavily attenuated, falling to 1/e of its initial amplitude in a distance

$$\delta = \frac{1}{k_I} = \left(\frac{2}{\mu_r\mu_0\omega\gamma}\right)^{1/2} \tag{12.13}$$

The distance δ is called the *skin depth* and is much less than the wavelength, $2\pi k$ of the electromagnetic wave in the conductor, as shown in Fig. 12.1. (It should not be confused with the penetration depth, Λ, in a superconductor, which refers to the decay of a static magnetic field, the Meissner effect.)

The magnetic vector of the linearly-polarized plane wave can be found by substituting E_x from (12.8) in the third Maxwell equation (12.4) to give

$$\begin{vmatrix} \hat{\mathbf{i}} & \hat{\mathbf{j}} & \hat{\mathbf{k}} \\ 0 & 0 & -\mathrm{i}k \\ E_x & 0 & 0 \end{vmatrix} = -\mathrm{i}\omega(B_x\hat{\mathbf{i}} + B_y\hat{\mathbf{j}} + B_z\hat{\mathbf{k}})$$

Hence,

$$B_y = \frac{k}{\omega}E_x \quad \text{or} \quad H_y = \left(\frac{k}{\mu_r\mu_0\omega}\right)E_x \tag{12.14}$$

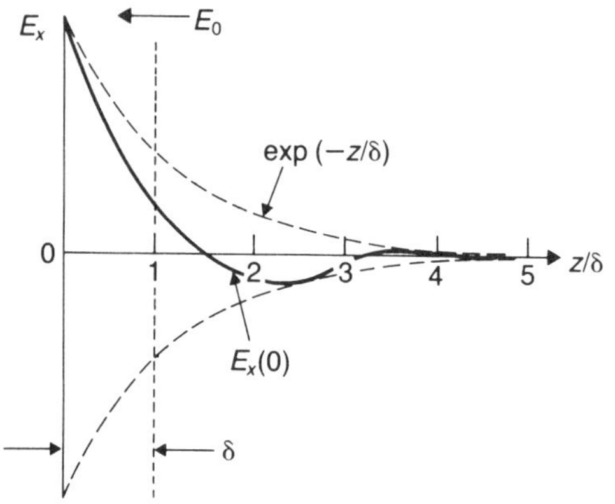

Fig. 12.1 An electromagnetic wave is heavily attenuated in a good conductor, penetrating less than a wavelength $(2\pi k)$, where δ is the skin depth given by equation (12.13).

where k is given by (12.12) in a good conductor. For this case we obtain

$$\frac{E_x}{H_y} = \frac{\mu_r \mu_0 \omega}{k} = \left(\frac{\mu_r \mu_0 \omega}{\gamma}\right)^{1/2} \exp \mathrm{i}(\pi/4) \tag{12.15}$$

showing that E leads H by 45° in good conductors, in contrast to free space and dielectrics where E and H are in phase (Fig. 9.6). This difference arises from the dominance of the conduction-current terms in (12.6) and (12.7) over the displacement-current terms in good conductors. In terms of the skin depth the electric and magnetic vectors are therefore

$$E_x = E_0 \exp \mathrm{i}(\omega t - z/\delta) \exp(-z/\delta) \tag{12.16}$$

$$H_y = \left(\frac{\gamma \delta E_0}{\sqrt{2}}\right) \exp \mathrm{i}\left(\omega t - \frac{z}{\delta} - \frac{\pi}{4}\right) \exp\left(-\frac{z}{\delta}\right) \tag{12.17}$$

The skin depth for a good conductor like copper is given by $\delta = (2/\mu_0 \gamma \omega)^{1/2}$ and so becomes very small at microwave frequencies. For example, at 10 GHz in pure copper at 293 K, $\delta = 0.67\ \mu\text{m}$, falling to as little as 10 nm at 4 K, so that a thin plating of copper is all that is needed to absorb microwave radiation. At radio frequencies the resistance R_{rf} of a cylindrical wire is very different from its zero-frequency resistance R_0, since the r.f. fields only penetrate into a surface sheath of thickness about δ. Therefore the r.f. resistance is

$$R_{\mathrm{rf}} = \frac{l}{\gamma(2\pi r\delta)} = \frac{rR_0}{2\delta} \tag{12.18}$$

and hence much finer wires can be used to provide low-resistance leads, often in the form of a braid, at radio frequencies, at much less cost than the rods necessary for direct currents.

12.2 WAVE IMPEDANCE AND REFLECTANCE

The wave impedance of a good conductor, from (11.26), (12.12) and (12.15), is

$$Z = \frac{E_x}{H_y} = \frac{\mu_r\mu_0\omega}{k} = \left(\frac{2\mu_r\mu_0\omega}{\gamma}\right)^{1/2}\frac{1}{(1-\mathrm{i})}$$

Substituting for δ from (12.13) this becomes

$$Z = \frac{(1+\mathrm{i})}{\gamma\delta} \tag{12.19}$$

and has the value 0.025 $(1+\mathrm{i})\,\Omega$ for copper at 10 GHz. This small value of wave impedance shows that the electric field is much less than the magnetic field and so the electromagnetic energy in a good conductor is nearly all magnetic energy. Since the wave impedance is also much less than for free space, $Z_0 = 377\,\Omega$ (9.38), a microwave incident from air on to a metal is almost totally reflected.

The analysis of the reflected and transmitted waves formed when a plane wave is incident from a dielectric on to a conductor follows similar lines to those uses for dielectrics in the previous chapter. Since the wave number and the wave impedance are complex for conductors, the phase changes at the boundary, which for dielectrics are always 0 or π, now vary, so that in general a plane-polarized wave after reflection will be elliptically polarized. The simplest case is that for normal incidence where, from (11.50), the reflectance is:

$$R_0 = \left(\frac{Z_2 - Z_1}{Z_2 + Z_1}\right)^2$$

and, for an air to metal reflection, $Z_1 = \mu_0 c$ and $Z_2 = (1+\mathrm{i})\gamma\delta$, Hence,

$$R_0 = \left|\frac{(1+\mathrm{i}) - \mu_0 c\gamma\delta}{(1+\mathrm{i}) + \mu_0 c\gamma\delta}\right|^2$$

which can be simplified if we put $a = \mu_0 c\gamma\delta$, to give

$$R_0 = \left|\frac{(1-a)+\mathrm{i}}{(1+a)+\mathrm{i}}\right|^2 = \frac{2-2a+a^2}{2+2a+a^2}$$

For copper at 293 K, $a = 4\pi \times 10^{-7} \times 3 \times 10^8 \times 6 \times 10^7\delta = 2.3 \times 10^{10}\delta$, so that at microwave frequencies $a \gg 1$. Hence,

$$R_0 = \frac{1-(2/a)+(2/a^2)}{1+(2/a)+(2/a^2)} \simeq 1-(4/a)$$

giving

$$R_0 = 1 - \frac{4}{\mu_0 c\gamma\delta} \tag{12.20}$$

as the reflectance at normal incidence of microwave radiation on a good conductor.

The almost perfect reflectance of metals for electromagnetic radiation in classical theory is associated with their high absorption within a few skin depths (Fig. 12.1). This is an example of the general rule for radiation, that 'good absorbers are good reflectors' at a particular frequency, and comes from the large value of the imaginary $k_\mathrm{I} = 1/\delta$. Dried red ink can sometimes be seen to give a greenish metallic reflection, showing that it absorbs green light, reflects green and transmits red.

The real part of the wave number, k_R, from (12.12), is also large and so the phase velocity v of a radio wave or microwave in a metal is very small, since

$$v = \frac{\omega}{k_\mathrm{R}} = \left(\frac{2\omega}{\mu_\mathrm{r}\mu_0\omega}\right)^{1/2} \tag{12.21}$$

For example at 10 MHz in copper at 293 K, $v = 1290\,\mathrm{m\,s^{-1}}$, much less than the speed of sound in copper.

12.3 ENERGY FLOW AND RADIATION PRESSURE

We saw in section 11.3 that the Poynting vector was $\mathscr{S} = \boldsymbol{E} \times \boldsymbol{H}$ in any medium and that the energy density was $u = \frac{1}{2}(\varepsilon_\mathrm{r}\varepsilon_0 E^2 + \mu_\mathrm{r}\mu_0 H^2)$ in a linear, isotropic medium. The average Poynting vector in a dielectric at optical frequencies, where n (and hence k) are real, was given by (11.42) as

$$\langle \mathscr{S} \rangle = \tfrac{1}{2} E_0 H_0 \hat{\mathbf{k}}$$

In a conductor, where the wave number k is complex, the average Poynting vector can be obtained from (exercise 1)

$$\langle \mathscr{S} \rangle = \tfrac{1}{2}\mathrm{Re}(\boldsymbol{E} \times \boldsymbol{H}^*) \qquad (12.22)$$

where Re means 'real part of' and $\boldsymbol{H}^*$ is the complex conjugate of $\boldsymbol{H}$, obtained by substituting $-\mathrm{i}$ for i.

For a good conductor the electric and magnetic vectors of a plane wave are given by (12.16) and (12.17). The energy density is therefore

$$u = \tfrac{1}{2}\{\varepsilon_r\varepsilon_0 E_0^2 + \mu_r\mu_0(\gamma^2\delta^2 E_0^2/2)\}$$

or

$$u = \frac{E_0^2}{2}\left\{\varepsilon_r\varepsilon_0 + \frac{\gamma}{\omega}\right\} \qquad (12.23)$$

from the definition of δ in (12.13). Hence the ratio of the magnetic to electric energy is $\gamma/(\omega\varepsilon_r\varepsilon_0)$ and this can be very large, e.g. 10^{11} for 10 MHz waves in copper at 293 K. Similarly, the average Poynting vector (12.22) is

$$\langle \mathscr{S} \rangle = \frac{1}{2}\left(\frac{\gamma}{2\omega\mu_r\mu_0}\right)^{1/2} \exp\left(-\frac{2z}{\delta}\right) E_0^2 \hat{\mathbf{k}} \qquad (12.24)$$

For a radio wave with $E_0 = 0.1\ \mathrm{V\,m^{-1}}$ in space, which has an intensity of $30\ \mu\mathrm{W\,m^{-2}}$ from (9.34), the intensity in a metal is much greater and at a depth of δ in copper at 10 MHz is about $5\ \mathrm{W\,m^{-2}}$.

An electromagnetic wave incident on a good conductor from a vacuum exerts a small radiation pressure. This can be deduced from the incident energy density u_i and energy flux $\mathscr{S}_i$ given by (9.35) and (9.34):

$$u_i = \frac{1}{2}\frac{E_i^2}{\mu_0 c^2} + \frac{1}{2}\varepsilon_0 E_i^2 = \varepsilon_0 E_i^2 \qquad (12.25)$$

$$\mathscr{S}_i = \varepsilon_0 c E_i^2 = c u_i \qquad (12.26)$$

At normal incidence there is near perfect *specular reflection* from a plane surface of a good conductor, so that the incident energy flux is totally reversed, giving rise to the radiation pressure, p_r, in the direction of the incident Poynting vector. The energy density is equivalent to a momentum p per unit volume at speed c across unit area or

$$u_i = pc$$

which gives a total change of momentum flux $= 2pc$ after reflection and hence a radiation pressure

$$p_r = 2pc = 2u_i = \frac{2\mathscr{S}_i}{c} \qquad (12.27)$$

If the radiation is diffuse, then on average one-third of the total energy density is associated with normal incidence and so, for *diffuse reflection*,

$$p_r = \frac{2}{3}u_i = \frac{2\mathscr{S}_i}{3c} \qquad (12.28)$$

The radiation pressure can also be calculated from the Lorentz force due to the magnetic vector acting on the induced surface current in a direction normal to the surface (exercise 2). It is extremely small for sunlight and even for an intense source such as a laser beam, where an intensity of $100\,\mathrm{GW\,m^{-2}}$ produces a radiation pressure of 670 Pa when specularly reflected, it is less than one-hundredth of an atmosphere. However, it is of vital importance in stars, where it prevents a gravitational collapse.

12.4 PLASMAS

A plasma is an electrically neutral, ionized gas consisting of equal numbers of light electrons and heavy ions. Since the ions have masses at least $m_p = 1836\ m_e$ they are assumed to be stationary and only the electrons are mobile. Here we consider only cold plasmas, that is we neglect the thermal motions of the electrons and ions. Since the simplest model of a metal is the free electron gas, in which the lattice of positive ion cores provides electrical neutrality, a metal can also be regarded as a plasma in its interaction with electromagnetic radiation.

The current density for N electrons per unit volume each of charge $-e$ and moving with velocity $\boldsymbol{v}_e$ is, from (4.1),

$$\boldsymbol{j} = -Ne\boldsymbol{v}_e \qquad (12.29)$$

while their equation of motion in the presence of the electromagnetic radiation is, from (4.3) and (4.9),

$$m\frac{d\boldsymbol{v}_e}{dt} = -e(\boldsymbol{E} + \boldsymbol{v}_e \times \boldsymbol{B}) - \frac{m\boldsymbol{v}_e}{\tau} \qquad (12.30)$$

where the Lorentz force is reduced by the momentum loss in collisions with a mean time τ between collisions. The magnetic force can be neglected since, from (12.14),

$$\frac{|\boldsymbol{v}_e \times \boldsymbol{B}|}{\boldsymbol{E}} \sim \frac{v_e k}{\omega} \sim \frac{v_e}{v}$$

and the phase velocity of the wave $\boldsymbol{v}$ is much greater than the electron velocity v_e. Assuming a frequency dependence exp $i\omega t$ for both $\boldsymbol{E}$ and $\boldsymbol{v}_e$, (12.30) becomes

$$\boldsymbol{v}_e = \frac{-e\boldsymbol{E}}{m(i\omega + 1/\tau)}$$

and so the current density from (12.29) is

$$\boldsymbol{j} = \frac{Ne^2\boldsymbol{E}\tau}{m(1 + i\omega\tau)} \tag{12.31}$$

The wave parameters of the plasma can be described, like those of a dielectric, in terms of a complex refractive index, $n = n_R - in_I$, (10.29), or a complex permittivity, $\varepsilon_r = \varepsilon_R - i\varepsilon_I$, (10.32). The latter follows from Maxwell's fourth equation (7.22)

$$\text{curl}\,\boldsymbol{H} = \boldsymbol{j}_f + \frac{\partial \boldsymbol{D}}{\partial t}$$

where in a plasma there is no polarization and so, using (12.31),

$$\frac{Ne^2\boldsymbol{E}\tau}{m(1 + i\omega\tau)} + i\omega\varepsilon_0\boldsymbol{E} = i\omega\varepsilon_r\varepsilon_0\boldsymbol{E}$$

or

$$\varepsilon_r = 1 + \frac{Ne^2\tau}{i\omega m\varepsilon_0(1 + i\omega\tau)} \tag{12.32}$$

At *low frequencies* $\omega\tau \ll 1$ and, in a good conductor, $\omega \ll \gamma/\varepsilon_0$, where $\gamma = Ne^2\tau/m$, so that we have

$$\varepsilon_r = -\frac{i\gamma}{\omega\varepsilon_0} \tag{12.33}$$

For copper at 293 K, $\gamma = 5.8 \times 10^7\,\Omega^{-1}\,\text{m}^{-1}$, $N = 8.5 \times 10^{28}$ so that $\tau = 2.4 \times 10^{-14}$ s and the limit for $\omega\tau \ll 1$ is a frequency less than 1 THz (millimetre waves). Since the skin depth $\delta = 1/k_I = c/\omega\varepsilon_r^{1/2}$ we recover the previous formula $\delta = (2/\mu_0\gamma\omega)^{1/2}$ for a good, paramagnetic conductor like copper.

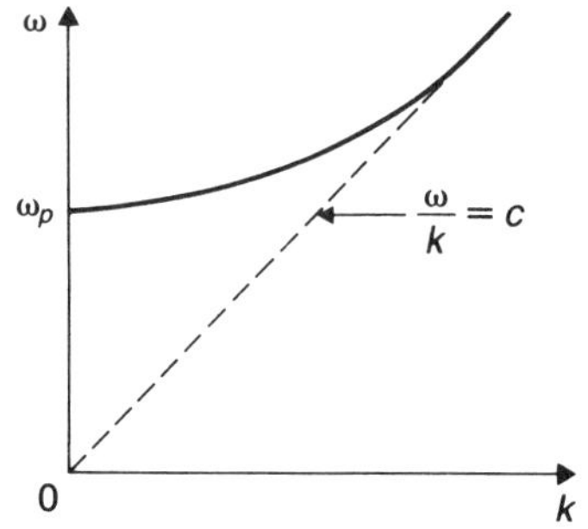

Fig. 12.2 The dispersion relation for high frequency ($\omega\tau \gg 1$) propagation in a plasma with plasma frequency ω_p.

At *high frequencies* $\omega\tau \gg 1$ and so

$$\varepsilon_r = 1 - \frac{Ne^2}{m\varepsilon_0\omega^2} \tag{12.34}$$

is a real permittivity, giving a real refractive index, which is less than one (as already seen for X-rays in dielectrics, (10.37). The *plasma frequency* ω_p is given by

$$\omega_p^2 = Ne^2/m\varepsilon_0 \tag{12.35}$$

so that (12.34) can also be written

$$\varepsilon_r = n^2 = \frac{k^2c^2}{\omega^2} = 1 - \frac{\omega_p^2}{\omega^2}$$

or

$$\omega^2 = \omega_p^2 + k^2c^2 \tag{12.36}$$

This is the dispersion relation for high-frequency propagation in a plasma and is shown in Fig. 12.2. Obviously for $\omega > \omega_p$ there is no attenuation and so a metal at these frequencies becomes transparent. For an alkali metal like sodium, absorption begins in the ultraviolet, but for less-dense plasmas (Table 12.1) the cut-off frequencies are much lower. For example in a gas discharge or a semiconductor with $N = 10^{21}\,\text{m}^{-3}$ the plasma frequency is 300 GHz, while in the ionosphere, with $N = 10^{11}\,\text{m}^{-3}$, it is only 3 MHz. In the latter case, since the electron density increases with height, medium radio waves ($f < 3$ MHz) are bent back to earth, providing long-distance terrestrial communications, while short waves ($f > 3$ MHz) are necessary for transmissions to satellites beyond the ionosphere.

13

Generation of electromagnetic waves

In the previous four chapters we have studied the propagation of electromagnetic waves in free space, in dielectrics and in conductors, while ignoring the question of their generation. Now we go back to the *inhomogeneous wave equations* and their solutions the *retarded vector and scalar potentials*. We first develop these potentials for the radiant energy a long way from an oscillating electric dipole and then discuss how the radiators of electromagnetic energy at radio and microwave wavelengths (antennas) can be made directional. Finally we consider the classical scattering of electromagnetic waves.

13.1 HERTZIAN DIPOLE

We saw in Chapter 8 that when we solved the inhomogeneous wave equations (8.33) and (8.32), for a distribution of moving charges and currents (Fig. 8.11), we obtained potentials at fixed field points $[1, t] \equiv (x, y, z, t)$ due to the charges and currents at the source points at the earlier time $(t - r_{12}/c)$. In this way the retarded potentials allow for the finite time taken to propagate at speed c, according to (8.52) and (8.53), respectively:

$$A(1,t) = \frac{\mu_0}{4\pi}\int \frac{j(2, t - r_{12}/c)}{r_{12}}\mathrm{d}\tau_1 \qquad (8.52)$$

$$\phi(1,t) = \frac{1}{4\pi\varepsilon_0}\int \frac{\rho(2, t - r_{12}/c)}{r_{12}}\mathrm{d}\tau_2 \qquad (8.53)$$

Having found A and ϕ for a particular source of radiation, the electric and magnetic fields, from (8.29) and (8.26), respectively, are given by:

$$\boldsymbol{E} = -\operatorname{grad}\phi - \frac{\partial \boldsymbol{A}}{\partial t}$$

$$\boldsymbol{B} = \operatorname{curl} \boldsymbol{A}$$

In general the calculation of $\boldsymbol{A}$, ϕ is complicated for finite sources and so we solve a simple case, the Hertzian (or oscillating) electric dipole. This is an approximation to an oscillator, angular frequency ω, connected to a dipole antenna length l (Fig. 13.1(a)) when $l \ll \lambda$, the wavelength of the radiation.

13.1.1 Potentials

Consider an electric dipole $\boldsymbol{p} = q\boldsymbol{l}$ along the z-axis and at the origin of a spherical polar coordinate system, Fig. 13.1(b). For an oscillating dipole moment, we have

$$q = q_0 \sin \omega t, \qquad \boldsymbol{p} = \boldsymbol{p}_0 \sin \omega t$$

but for points at a distance $r \gg l$, we may neglect the time taken for a signal to traverse the source $(T/2)$ compared with the time (r/c) for propagation to the field point $\boldsymbol{r}$. The current I, given by

$$I = \frac{\mathrm{d}q}{\mathrm{d}t} = I_0 \cos \omega t \tag{13.1}$$

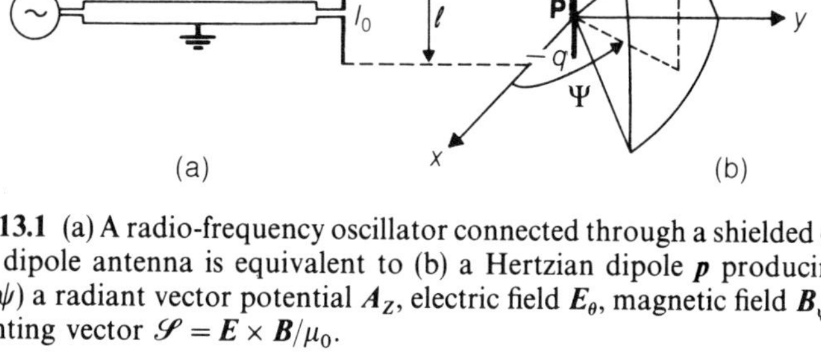

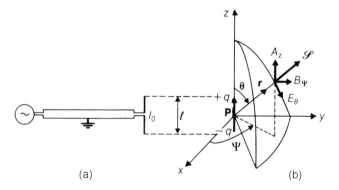

Fig. 13.1 (a) A radio-frequency oscillator connected through a shielded cable to a dipole antenna is equivalent to (b) a Hertzian dipole $\boldsymbol{p}$ producing at (r, θ, ψ) a radiant vector potential $\boldsymbol{A}_Z$, electric field $\boldsymbol{E}_\theta$, magnetic field $\boldsymbol{B}_\psi$ and Poynting vector $\mathscr{S} = \boldsymbol{E} \times \boldsymbol{B}/\mu_0$.

where $I_0 = \omega q_0$ is constant across l, is related to the current density $\boldsymbol{j}$ for a thin wire by

$$\boldsymbol{j}\,\mathrm{d}\tau = I\mathbf{d}\boldsymbol{l}$$

where $\mathbf{d}\boldsymbol{l}$ is a vector in the direction of the current, here along Oz. Therefore the dipole $\boldsymbol{p}$ generates only the z component of the vector potential $\boldsymbol{A}$ and (8.52) becomes

$$A_z(1,\ t) = \frac{\mu_0}{4\pi}\int I\frac{(2, t - r_{12}/c)}{r_{12}}\mathrm{d}l$$

For a distant point ($r \gg l$) of a dipole ($\lambda \gg l$) we may put $r_{12} = |\boldsymbol{r}_1 - \boldsymbol{r}_2| \simeq r$ and take the integration of the current from $-l/2$ to $+l/2$, so that

$$A_z(r,\ t) = \left(\frac{\mu_0 l}{4\pi}\right)\frac{I(t - r/c)}{r} \tag{13.2}$$

Thus $\boldsymbol{A}$ is everywhere parallel to $\boldsymbol{p}$, Fig. 13.1(b), and decreases as $1/r$, as expected for a spherical wave, see (9.23).

The scalar potential can be found most easily from the Lorentz condition (8.31),

$$\operatorname{div}\boldsymbol{A} = -\frac{1}{c^2}\frac{\partial\phi}{\partial t}$$

where here

$$\operatorname{div}\boldsymbol{A} = \frac{\partial A_z}{\partial z} = \left(\frac{\mu_0 l}{4\pi}\right)\frac{\partial}{\partial z}\left\{\frac{I(t - r/c)}{r}\right\} \tag{13.3}$$

The solution for ϕ, after differentiating the product and integrating the time derivative (exercise 1), is

$$\phi = \frac{l}{4\pi\varepsilon_0}\left\{\frac{\cos\theta}{r^2}\cdot q(t - r/c) + \frac{\cos\theta}{cr}I(t - r/c)\right\} \tag{13.4}$$

where $q(t - r/c) = q_0 \sin\omega(t - r/c)$ and $I(t - r/c) = \omega q_0 \cos\omega(t - r/c)$.

13.1.2 Fields

From (8.26) and (13.2), the magnetic field

$$\boldsymbol{B} = \operatorname{curl}\boldsymbol{A} = \frac{\mu_0}{4\pi}\operatorname{curl}\left\{\frac{I(t - r/c)}{r}\boldsymbol{l}\right\}$$

and, using the vector identity for the curl of a product (Appendix E), we have

$$\boldsymbol{B} = \frac{\mu_0}{4\pi}\left[\frac{I(t-r/c)}{r}\operatorname{curl}\boldsymbol{l} + \operatorname{grad}\left\{\frac{I(t-r/c)}{r}\right\} \times \boldsymbol{l}\right]$$

But curl $\boldsymbol{l}=0$, since $\boldsymbol{l}$ is along Oz, and I varies in space only with $\boldsymbol{r}$, so that

$$\boldsymbol{B} = \frac{\mu_0}{4\pi}\left[\boldsymbol{l} \times \frac{\partial}{\partial r}\left\{\frac{I_0 \cos\omega(t-r/c)}{r}\right\}\hat{\mathbf{r}}\right] \tag{13.5}$$

Therefore $\boldsymbol{B}$ is normal to both $\boldsymbol{l}$ and $\boldsymbol{r}$ and so must be tangential to an azimuthal circle, that is, $\boldsymbol{B} = B_\psi \hat{\boldsymbol{\psi}}$. Since $\boldsymbol{l} \times \hat{\mathbf{r}} = l\sin\theta$, (13.5) becomes:

$$B_\psi = \left(\frac{\mu_0 l I_0 \sin\theta}{4\pi}\right)\frac{\cos\omega(t-r/c)}{r^2} - \left(\frac{\mu_0 \omega l I_0 \sin\theta}{4\pi c}\right)\frac{\sin\omega(t-r/c)}{r} \tag{13.6}$$

This expression shows the general result for the electromagnetic field due to changing currents. It consists of two terms:

(i) A term decreasing more rapidly than $1/r$—the *induction* (or near) *field*;
(ii) A term decreasing as $1/r$—the *radiation* (or far) field.

The induction field dominates when $r \ll \lambda$ and in this case is just the field given by the law of Biot and Savart, equation (8.44), that we found by taking curl $\boldsymbol{A}$ for a circuit $\oint I\,\mathbf{d}\boldsymbol{l}$. If we draw a sphere of large radius round the dipole the total energy into it is given by $\int \mathscr{S}\cdot\mathbf{d}\boldsymbol{S}$ over the surface of the sphere, so that the contribution to this integral of field terms that decrease more rapidly than $1/r$ tend to zero for a sphere of large enough radius. Hence in calculating the *radiant* energy we include only the *radiation field* term, which decreases as $1/r$.

In a similar way the electric radiation field E_θ is found in spherical polar coordinates from (8.29), (13.2) and (13.4) to be (exercise 2)

$$E_\theta = -\left(\frac{\omega l I_0 \sin\theta}{4\pi\varepsilon_0 c^2}\right)\frac{\sin\omega(t-r/c)}{r} \tag{13.7}$$

i.e. normal to $\hat{\boldsymbol{r}}$ and B_ψ, as shown in Fig. 13.1(b). In amplitude the

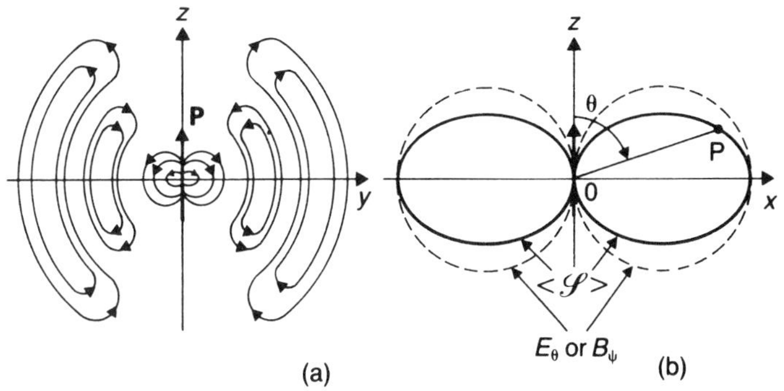

Fig. 13.2 (a) Electric field lines in the axial plane of the Hertzian dipole for $\omega t = 0, 2\pi, 4\pi\ldots$ The pattern is the same for $\omega t = \pi, 3\pi, 5\pi$, but with all the directions reversed. The magnetic field lines are azimuthal circles about the vertical axis of the dipole. (b) Polar plots of the amplitude of the electric $\boldsymbol{E}_\theta$ and magnetic $\boldsymbol{B}_\psi$ fields (----) and of average Poynting vector $\langle \mathscr{S} \rangle$ (——) for a Hertzian dipole. The radiant energy in the direction θ is proportional to the length OP and independent of the azimuthal angle ψ, so that in three dimensions the polar plot has a doughnut shape with no radiation along the axis and a maximum in the equatorial plane.

ratio

$$\frac{E_\theta}{B_\psi} = \frac{1}{\mu_0 \varepsilon_0 c} = c \tag{13.8}$$

for spherical waves in space, as was found for plane waves (9.18). The electric-field lines in the axial plane are plotted in Fig. 13.2(a); they have cylindrical symmetry about the dipole axis. Both the electric and magnetic vectors vary as $\sin\theta$ in amplitude, as shown in Fig. 13.2(b) by the dashed circles.

13.2 RADIANT ENERGY AND POWER

The radiant energy crossing unit area per second, the energy flux (or Poynting) vector (9.29), in free space, is:

$$\mathscr{S} = \boldsymbol{E} \times \boldsymbol{B}/\mu_0$$

which gives for the Hertzian dipole

$$\mathscr{S} = \left\{ -\frac{E_0}{r} \sin \omega(t - r/c)\hat{\boldsymbol{\theta}} \right\} \times -\left\{ \frac{E_0}{\mu_0 c r} \sin \omega(t - r/c)\,\hat{\boldsymbol{\psi}} \right\}$$

where $E_0 = (\omega l I_0 \sin\theta)/(4\pi\varepsilon_0 c^2)$. Since $\hat{\boldsymbol{\theta}} \times \hat{\boldsymbol{\psi}} = \hat{\mathbf{r}}$, $\mathscr{S}$ is a radial vector, (Fig. 13.1(b)) as expected for a spherical wave, and

$$\mathscr{S} = \left(\frac{E_0^2}{\mu_0 c r^2}\right)\sin^2\omega(t - r/c)\hat{\mathbf{r}} \tag{13.9}$$

The time-averaged Poynting vector, therefore (exercise 2), is

$$\langle\mathscr{S}\rangle = \left(\frac{\mu_0 c l^2}{32\pi^2}\right) I_0^2 \sin^2\theta \frac{k^2}{r^2}\hat{\mathbf{r}} \tag{13.10}$$

where $k = \omega/c$ is the wave number for the radiation. Since $\langle\mathscr{S}\rangle$ is axially symmetrical (independent of ψ) it is usually drawn on a polar plot (Fig. 13.2(b)), which is a vertical section of a doughnut-shaped surface. Its polar variation is as $\sin^2\theta$, so that there is no radiation along the axis of the dipole ($\theta = 0$) and a maximum output in the equatorial plane ($\theta = \pi/2$). As expected $\langle\mathscr{S}\rangle$ is proportional to $1/r^2$ so that the energy flow through any solid angle is the same for all r. Since $\langle\mathscr{S}\rangle$ varies as $l^2 I_0^2 k^2$, which is proportional to $\omega^4 p_0^2$, radiation is much more efficient at high source frequencies.

The total radiated power P is obtained by integrating $\langle\mathscr{S}\rangle$ over the surface of a sphere of radius $\mathbf{r}$, that is

$$P = \int_S \langle\mathscr{S}\rangle\cdot\mathbf{dS}$$

where $\mathrm{d}S = r\sin\theta\,\mathrm{d}\psi r\mathrm{d}\theta$.

Hence

$$P = \left(\frac{\mu_0 c l^2 I_0^2 k^2}{32\pi^2}\right)\int_0^\pi \left(\frac{\sin^2\theta}{r^2}\right) r^2 \sin\theta\,\mathrm{d}\theta \int_0^{2\pi}\mathrm{d}\psi$$

or

$$P = (\mu_0 c l^2 I_0^2 k^2)/12\pi \tag{13.11}$$

This power is conveniently expressed as

$$P = \tfrac{1}{2} R_r I_0^2 \tag{13.12}$$

where R_r is the *radiation resistance* of the source, since this is just the mean power that would be dissipated in a resister whose $I_{\mathrm{rms}}^2 = I_0^2/2$. For the Hertzian dipole

$$R_r = \frac{\mu_0 c l^2 k^2}{6\pi} = 20(kl)^2 \tag{13.13}$$

and so depends only on the ratio l/λ, which we have assumed to be small. Thus a dipole antenna can be matched quite easily to a low impedance shielded cable (Fig. 13.1(a)).

13.3 ANTENNAS

The principles that we have used to calculate the electromagnetic field, the radiated power and the radiation resistance for the Hertzian dipole can also be applied to other sources of radiation, such as microscopic sources (magnetic dipoles and electric quadrupoles) and macroscopic sources (radio antennas).

13.3.1 Microscopic sources

A magnetic dipole in the form of a current loop of area $dS = \pi a^2$ has magnetic moment,

$$\boldsymbol{m}_0 = I\mathbf{d}\boldsymbol{S} \tag{13.14}$$

where $\boldsymbol{m}_0$ is along Oz for a current loop in the xy plane. For an oscillating dipole moment

$$m = m_0 \cos \omega t \tag{13.15}$$

the electromagnetic field can be shown to be similar to that for the electric dipole, Fig. 13.1(b), expect that B_ψ is now replaced by E_ψ and E_θ is replaced by $-B_\theta$, thus keeping $\mathscr{S}$ along $\boldsymbol{r}$ at a distance point $r \gg a$ for wavelengths $\lambda \gg a$. For a small loop the scalar potential ϕ will be constant and can be set to zero, so that both of the components of the electromagnetic field are due to the magnetic vector potential $\boldsymbol{A}$. They can be shown to be

$$E_\psi = \left(\frac{\mu_0 \omega^2 m_0 \sin\theta}{4\pi c}\right)\frac{\cos\omega(t - r/c)}{r} \tag{13.16}$$

and

$$-B_\theta = \left(\frac{\mu_0 \omega^2 m_0 \sin\theta}{4\pi c^2}\right)\frac{\cos\omega(t - r/c)}{r} \tag{13.17}$$

so that again $|\boldsymbol{E}/\boldsymbol{B}| = c$ and the time-averaged Poynting vector is

$$\langle \mathscr{S} \rangle = \left(\frac{\mu_0 c m_0^2}{32\pi^2}\right)\sin^2\theta\left(\frac{k^4}{r^2}\right)\hat{\mathbf{r}} \tag{13.18}$$

Comparing this with (13.10) we see that $\langle \mathscr{S} \rangle$ is now proportional to $\omega^4 m_0^2$ rather than $\omega^4 p_0^2$. The corresponding radiation resistance (exercise 4) is $20\pi^2(ka)^4$.

A linear electric quadrupole can be formed by placing two dipoles end-on so that their negative charges coincide, Fig. 13.3(a). It has a quadrupole moment $Q_{zz} = \sum_i q z_i^2 = 2ql^2$ and if each charge q oscillates in amplitude as $q_0 \cos \omega t$, then the quadrupole moment is

$$Q_{zz} = Q_0 \cos \omega t \tag{13.19}$$

where $Q_0 = 2q_0 l^2$. Considering only the radiation field, we can find its field components by superimposing the dipole fields of electric dipoles centred at $z = \pm l/2$, after allowing for their phase difference. The resultant fields are

$$E_\theta = \left(-\frac{\mu_0 \omega^3 Q_0 \sin\theta \cos\theta}{8\pi c} \right) \frac{\cos \omega(t - r/c)}{r} \tag{13.20}$$

$$B_\psi = -\left(\frac{\mu_0 \omega^3 Q_0 \sin\theta \cos\theta}{8\pi c^2} \right) \frac{\cos \omega(t - r/c)}{r} \tag{13.21}$$

There can be no radiation along the directions $\theta = 0$, π, since these are the dipole axes, nor along $\theta = \pi/2$, where the dipole fields cancel. The field patterns are therefore given by the dashed curve in Fig. 13.3(b) and the polar diagram for $\langle \mathscr{S} \rangle$, which varies as $\sin^2\theta \cos^2\theta$, is the solid curve. It is noticeable that the fields now vary as $\omega^3 Q_0$

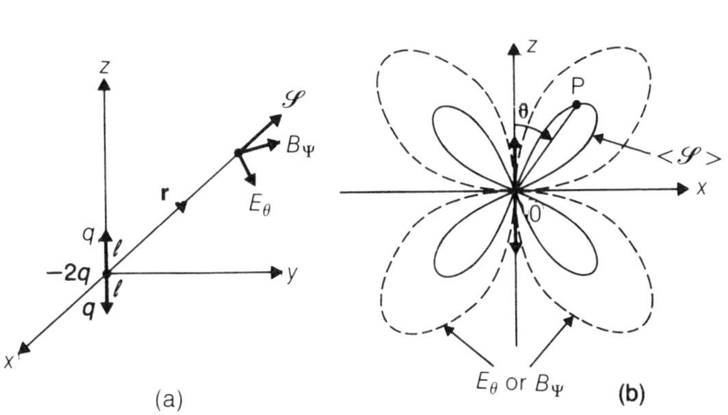

Fig. 13.3 (a) A linear electric quadrupole and its radiation field. (b) Polar plots of the amplitude of the electric and magnetic fields (----) and of the average Poynting vector (——) for a linear quadrupole. The plots are independent of ψ.

and the power as $\omega^6 Q_0^2$, giving a radiation resistance (exercise 5) of $4(kl)^4$.

13.3.2 Macroscopic sources

Most sources of radio waves use a half-wave ($\lambda/2$) antenna, Fig. 13.4(a), or a combination of $\lambda/2$ antennas. In the $\lambda/2$ antenna the current is

$$I = I_0 \cos(kl) \cos \omega t \tag{13.22}$$

where l is now a variable and each current element $I\,\mathbf{d}\boldsymbol{l}$ acts as an electric dipole producing a radiation field, from (13.7), of

$$\mathrm{d}E_\theta = -\left(\frac{\omega\,\mathrm{d}l I_0 \cos(kl)\sin\theta'}{4\pi\varepsilon_0 c^2}\right)\frac{\sin\omega(t - r'/c)}{r'} \tag{13.23}$$

where $\boldsymbol{r}'$, θ' are shown on Fig. 13.4(a). The total field is given by the coherent superposition of these dipole fields with $r' \approx r - l\cos\theta$, $\sin\theta'/r' \approx \sin\theta/r$, but with the phase differences retained, so that

$$E_\theta = -\frac{\omega I_0}{4\pi\varepsilon_0 c^2}\int_{-\lambda/4}^{\lambda/4}\frac{\cos(kl)\sin\theta\sin\omega(t - r'/c)}{r}\,dl$$

Calculation shows that

$$E_\theta = \frac{I_0}{2\pi\varepsilon_0 c}\cdot\frac{\cos\left(\dfrac{\pi}{2}\cos\theta\right)}{\sin\theta}\cdot\sin\omega(t - r/c) \tag{13.24}$$

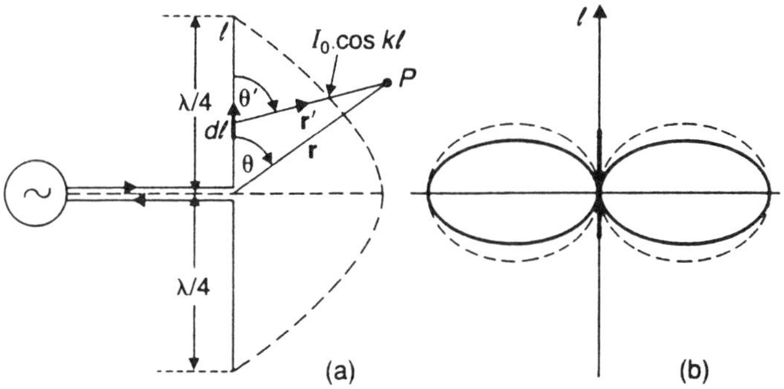

Fig. 13.4 (a) Half-wave antenna and its current distribution at $t = 0$. (b) Polar diagrams of the radiation fields (------) and radiant energy flux from a half-wave antenna.

and that

$$B_\psi = E_0/c \tag{13.25}$$

as with the dipole. The resultant field and energy-flux polar diagram, Fig. 13.4(b), is very similar to that of the dipole, Fig. 13.2(b), but the $\lambda/2$ antenna is slightly more directional. On the other hand the fields are now independent of the frequency for a given current, unlike the dipole fields, following the integration over $\lambda/2$.

A simple way to make a directional antenna is to place two $\lambda/2$ antennas $\lambda/4$ apart, Fig. 13.5(a), and to supply them with equal currents $\pi/2$ out of time phase. The coherent superposition of the radiation fields then produces almost perfect cancellation in one direction and an enhanced energy flux in the other, as shown for the plane normal to the antennas in Fig. 13.5(b). At higher frequencies (smaller wavelengths) a parabolic reflector with the source at its focus produces a parallel beam, Fig. 13.6(a), while at microwave frequencies a waveguide horn, Fig. 13.6(b), gives a highly directional beam.

Application to the *reciprocity theorem*, described in Pointon and Howarth 'AC and DC Network Theory' in this series, to a pair of antennas shows that the current in a receiving antenna divided by the voltage at the transmitting antenna remains constant when the source and detector are interchanged, provided all the impedances and frequency are constant. This is a valuable aid in designing antennas, since it means that the polar diagram of a transmitting antenna is the same as the polar response of the same antenna used as a receiver.

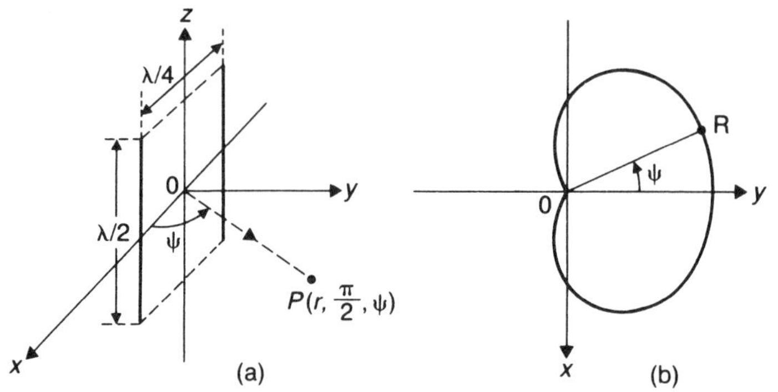

Fig. 13.5 (a) Two half-wave antenna spaced $\lambda/4$ apart and with currents $\pi/2$ out of phase. (b) Polar diagram of these antennas for radiant energy flux in the azimuthal plane, where the flux in direction ψ is proportional to OR.

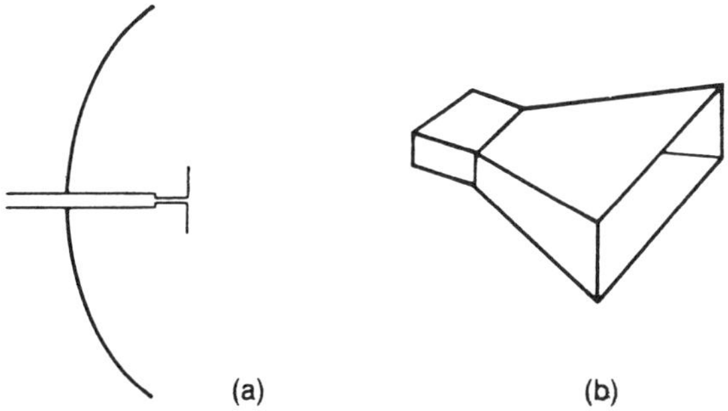

(a) (b)

Fig. 13.6 Directional antennas: (a) parabolic reflector with half-wave dipole at its focus to produce a parallel beam; (b) pyramidal horn ('cheese') antenna at end of a waveguide for a microwave beam.

13.4 SCATTERING

When electromagnetic radiation interacts with a charge distribution, such as a molecule, the resultant motion of the charges becomes a secondary source. This process is termed *scattering* of the incident radiation. We consider here the elastic scattering of the radiation from oscillating charges at non-relativistic speeds.

The total radiated power P of a Hertzian dipole, (13.11), can be expressed more generally (exercise 6) in terms of the second derivative of the dipole moment, evaluated at the retarded time $(t - r/c)$, as shown by the square brackets,

$$[\ddot{\boldsymbol{p}}] = \omega^2[\boldsymbol{p}] \tag{13.26}$$

as

$$P = [\ddot{\boldsymbol{p}}]^2/(6\pi\varepsilon_0 c^3) \tag{13.27}$$

which is known as *Larmor's formula*. We can apply this to both bound and free electrons.

For bound electrons we have an electric dipole polarizability $\alpha(\omega)$, (10.8), and a displacement x given by (10.4) and (10.6),

$$x = \frac{-eE_0 \exp(\mathrm{i}\omega t)}{m(\omega_0^2 - \omega^2 + \mathrm{i}\omega\gamma)} \tag{13.28}$$

where ω_0 is the natural frequency and γ the damping constant of the oscillating electrons. The instantaneous dipole moment is

$$\boldsymbol{p} = -e\boldsymbol{x} \tag{13.29}$$

and so, from (13.27), (13.28) and (13.29), the power radiated by the oscillating electrons is

$$P = \frac{e^4 E_0^2 \omega^2}{12\pi\varepsilon_0 m^2 c^3 \{(\omega_0^2 - \omega^2)^2 + \omega^2\gamma^2\}} \tag{13.30}$$

The *scattering cross-section*, σ, is the ratio of the radiated power to the power per unit area in the incident beam, from (9.34). That is,

$$\langle \mathscr{S} \rangle = \tfrac{1}{2}\varepsilon_0 c E_0^2 \tag{13.31}$$

Hence

$$\sigma = \frac{P}{\langle \mathscr{S} \rangle} = \frac{8\pi r_0^2}{3} \frac{\omega^4}{\{(\omega_0^2 - \omega^2)^2 + \omega^2\gamma^2\}} \tag{13.32}$$

where

$$r_0 = \frac{e^2}{4\pi\varepsilon_0 mc^2} \tag{13.33}$$

is the *classical electron radius*, about 2.8 fm.

The frequency dependence of this cross-section is shown in Fig. 13.7. At low frequencies, $\omega \ll \omega_0$, and since $\gamma \ll \omega_0$, the cross-section becomes

$$\sigma_R = \frac{8\pi r_0^2}{3}\left(\frac{\omega}{\omega_0}\right)^4 \tag{13.34}$$

which is the *Rayleigh scattering law*. It shows that scattering varies as λ^{-4} and so short wavelengths are scattered much more than long

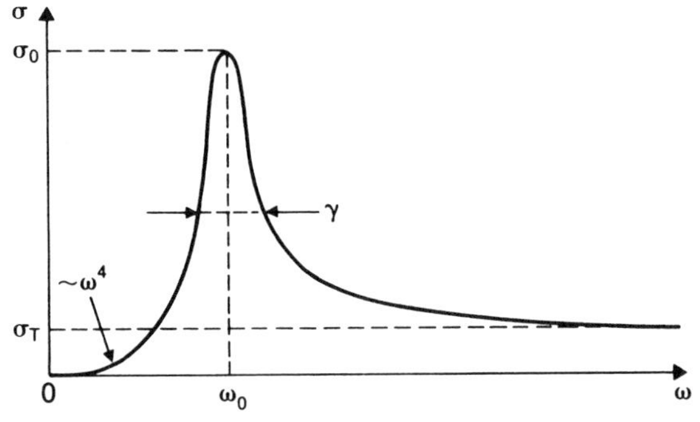

Fig. 13.7 Cross-section for elastic scattering of radiation by a molecule with electronic oscillators of natural frequency ω_0 and damping constant γ. For $\omega \ll \omega_0$ there is Rayleigh scattering and for $\omega \gg \omega_0$ there is Thomson scattering, cross-section σ_T.

wavelengths. A familiar example is the sky: blue in the day due to scattered sunlight, but red at sunrise and sunset when seen in the reflection of the sun's rays from high clouds.

When the frequency of the incident radiation is the same as the natural frequency of the oscillating electrons, the induced dipole moment is very large and the scattered radiation peaks at ω_0. This is known as *resonant scattering*, or resonance fluorescence in quantum theory, and produces a Lorentzian-shaped line of amplitude

$$\sigma_0 = \frac{8\pi r_0^2}{3}\left(\frac{\omega_0}{\omega}\right)^2 \tag{13.35}$$

At the highest frequencies, $\omega \gg \omega_0$, $\gamma \ll \omega_0$, the electron is effectively free, and its displacement becomes

$$x = \frac{eE_0}{m\omega^2}\exp(\mathrm{i}\omega t)$$

and the cross-section is just

$$\sigma_{\mathrm{T}} = \frac{8\pi r_0^2}{3} \tag{13.36}$$

This is *Thomson scattering* of free electrons and has a very small cross-section, from (13.33), of only $6.6 \times 10^{-29}\,\mathrm{m}^2$. Only electrons in a dense plasma are likely to produc significant Thomson scattering, one example being the corona seen round the sun as a bright ring during a total eclipse of the sun. In terms of σ_{T} we see that

$$\frac{\sigma_{\mathrm{R}}}{\sigma_{\mathrm{T}}} = \left(\frac{\omega}{\omega_0}\right)^4 \quad \text{and} \quad \frac{\sigma_0}{\sigma_{\mathrm{T}}} = \left(\frac{\omega_0}{\gamma}\right)^2 \tag{13.37}$$

At still higher frequencies classical theory no longer applies and we get photon-electron collisions or Compton scattering. The classical limit is $\hbar\omega \ll mc^2$ for Thomson scattering, from quantum theory.

14

Guided waves

In this final chapter we consider first the difference between free space and guided waves and then develop the theory for electromagnetic waves in rectangular waveguides. A discussion of waveguide modes is followed by an introduction to resonant cavities and the chapter concludes with the shortcomings of the classical theory of cavity radiation.

14.1 WAVEGUIDE EQUATION

A familiar experience when driving with a car radio on is to find the broadcast signal cut out within a short tunnel, although the end of the tunnel can be clearly seen in daylight. In such circumstances the tunnel is probably lined with steel-reinforced concrete and the mesh of steel wires forms a waveguide which evidently will not pass radio waves. Why should this waveguide pass electromagnetic waves at optical wavelengths but not at radio ones? An elegant answer to this question, due to Feynman, is to consider a line source S between two infinite, parallel plane conductors (Fig. 14.1(a)) spaced a apart. We know from the method of images in electrostatics that conductors act like mirrors in computing the resultant electric field, and the same is true for electrodynamics. The source S and the reflecting waveguide can therefore be replaced by a doubly infinite set of images $1, 1', 1'' \ldots, 2, 2', 2'' \ldots$ (Fig. 14.1(b)). If the initial phase of S is positive then the initial phase of the images 1, 2 will be negative, of the images 1′ and 2′ will be positive, etc., so that the fields are zero at the mirrors, that is at the conducting walls.

The resultant superposition of waves can be calculated from the *retarded* potentials, (8.52) and (8.53), and will produce both near and

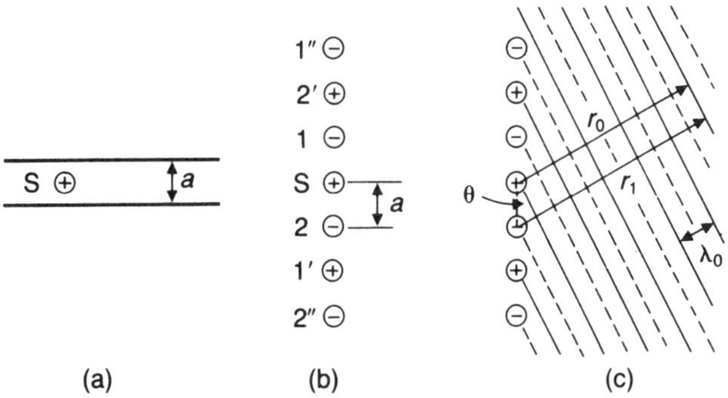

Fig. 14.1 (a) A line source S placed between two infinite parallel plane conductors is equivalent to (b) a source S plus multiple images $1, 1', 1'', \ldots$, $2, 2', 2'', \ldots$ in the conductors, which radiate (c) coherently in direction $(\theta + \pi/2)$.

far fields, as did the Hertzian dipole, (13.6). For the source S and its images these fields all cancel, but the far fields depend on the direction of propagation. For those directions in which all the fields are *in phase*, the fields will be strong and propagate as plane waves (Fig. 14.1(c)) so that

$$r_1 - r_0 = \lambda_0/2$$

and

$$\sin\theta = \lambda_0/2a \tag{14.1}$$

By symmetry, a similar set must propagate at $-\theta$ and when these two are superimposed we get the complete fields.

In Fig. 14.2 we have used the image construction to draw the plane waves within the waveguide and we see that the wave pattern goes from a double trough at P to a double crest at Q to a double trough at R, etc., at a phase velocity v. The *waveguide wavelength* is therefore λ_g, which is clearly longer than the free-space wavelength λ_0, since

$$\lambda_0 = \lambda_g \cos\theta \tag{14.2}$$

However, from (14.1)

$$\cos^2\theta = 1 - (\lambda_0/2a)^2 \tag{14.3}$$

so that

$$\frac{1}{\lambda_0^2} = \frac{1}{\lambda_g^2} + \frac{1}{(2a)^2} \tag{14.4}$$

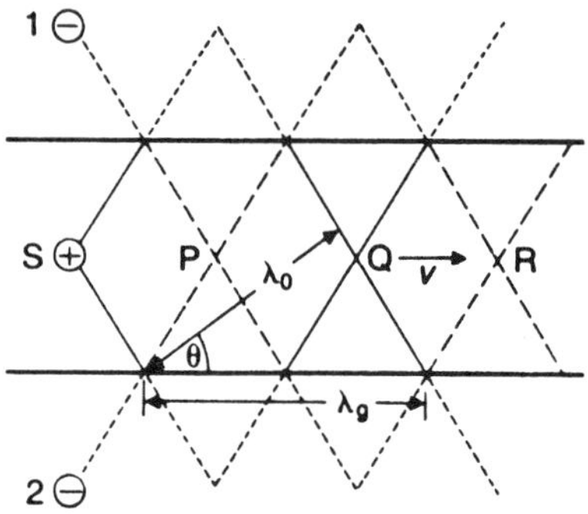

Fig. 14.2 The wave in the waveguide is formed by the superposition of the coherent radiations from the image sources and has a wavelength $\lambda_g > \lambda_0$, the free space wavelength.

Now we can see that if λ_0 is a long wavelength, greater than $2a$, it is impossible to find an angle θ for propagation. Therefore the constructive interference of the reflected waves, which allows the signal to travel down the guide, only occurs for wavelengths $\lambda_0 < 2a$. For this waveguide of two infinite conducting planes, the *cut-off wavelength* $\lambda_c = 2a$ and (14.4) becomes

$$\frac{1}{\lambda_0^2} = \frac{1}{\lambda_g^2} + \frac{1}{\lambda_c^2} \tag{14.5}$$

which is known as the waveguide equation.

14.2 RECTANGULAR WAVEGUIDES

For microwaves and millimetre waves the common shape for a waveguide is rectangular. It also has the simplest equations for its electric and magnetic fields, which we will now derive.

14.2.1 Electric field

A typical waveguide is shown in Fig. 14.3(a), where the widths are $x = a$, $y = b$ with $a \approx 2b$, and the wave propagates along z. The field vectors $\boldsymbol{E}$ and $\boldsymbol{B}$ of this wave must satisfy Maxwell's equations, but there are many possible solutions, or *waveguide modes*, for propagating waves that satisfy the boundary conditions (see section 11.1). The simplest mode for this rectangular waveguide is a wave with the electric field everywhere transverse (E_y): it is a TE mode. In lowest order it must be zero at $x = 0$ and $x = a$, since there is no tangential

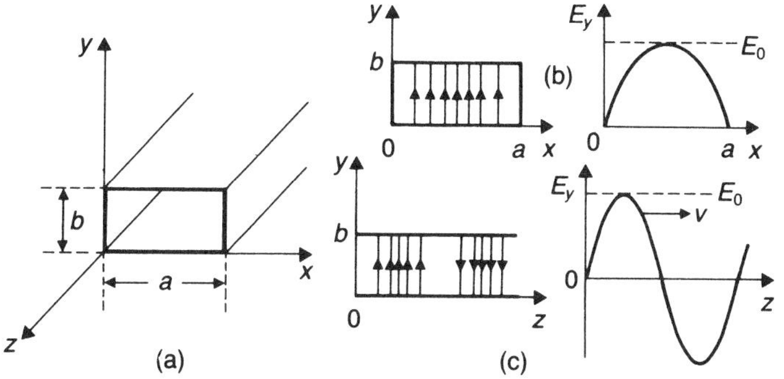

Fig. 14.3 (a) Rectangular waveguide, $x = a$, $y = b$. (b) Simple transverse electric (TE) mode. (c) Wave travels with phase velocity v along the waveguide (at $t = 2\pi/4\omega$).

component of $\boldsymbol{E}$ at the surface of a conductor, and will have a maximum in the centre (Fig. 14.3(b)) so that

$$E_y = E_0 \sin k_x x \tag{14.6}$$

where $k_x = 2\pi/2a$. If the wave travels with a phase velocity $v = \omega/k_g$, where $k_g = 2\pi/\lambda_g$, then, as in Fig. 14.3(c)

$$E_y = E_0 \sin k_x x \exp \mathrm{i}(\omega t - k_g z) \tag{14.7}$$

where k_g must satisfy Maxwell's equations.

Since $\partial E_y/\partial y = 0$, E_y satisfies div $\boldsymbol{E} = 0$, equation (7.5). The other Maxwell equations lead to (section 9.1)

$$\nabla^2 \boldsymbol{E} = \frac{1}{c^2}\frac{\partial^2 \boldsymbol{E}}{\partial t^2} \tag{14.8}$$

where here $\boldsymbol{E} = 0\hat{\mathbf{i}} + E_y\hat{\mathbf{j}} + 0\hat{\mathbf{k}}$. From (14.6) and (14.7),

$$\nabla^2 \boldsymbol{E} = -k_x^2 E_y + 0 - k_g^2 E_y$$

and

$$\partial^2 E_y/\partial t^2 = -\omega^2 E_y$$

so that

$$k_x^2 + k_g^2 = \omega^2/c^2 \tag{14.9}$$

Since $k_x = 2\pi/2a$ for this fundamental mode, this can be rearranged as

$$\frac{1}{(2a)^2} + \frac{1}{\lambda_g^2} = \frac{1}{\lambda_0^2}$$

and we recover (14.4) for the guide wavelength.

14.2.2 Attenuated wave

In section 14.1 we found that λ_0 had a maximum value $\lambda_c = 2a$, the cut-off wavelength. If we try and propagate a wave with $\lambda_0 > \lambda_c$, or $\omega_0 < \pi c/a$, then (14.9) shows that k_g becomes imaginary, as we found for a wave travelling into a conductor (section 12.1). Evidently the wave decays in amplitude as it enters the guide and for $\lambda_0 \gg \lambda_c$ this happens quite rapidly (Fig. 14.4) where $k_g = ik_I$ and (14.7) becomes

$$E_y = E_0 \sin k_x x \exp i(\omega t - k_I z) \tag{14.10}$$

From (14.9) when $\lambda_0 \gg \lambda_c = 2a$, $k_I = \pi/a$ and the wave penetrates only a distance a/π (Fig. 14.4) before its amplitude falls to E_0/e.

14.2.3 Travelling wave

The phase velocity of a propagating wave is, from (14.9),

$$v = \frac{\omega}{k_g} = \frac{1}{\left(\dfrac{1}{c^2} - \dfrac{k_x^2}{\omega^2}\right)^{1/2}}$$

and for the lowest mode, where $k_x = 2\pi/2a = 2\pi/\lambda_c = \omega_c/c$,

$$v = c/\{1 - (\omega_c/\omega)^2\}^{1/2} \tag{14.11}$$

For a travelling wave $\lambda < \lambda_c$ or $\omega > \omega_c$, so the *phase velocity is always greater than c*. However, the group velocity $u = d\omega/dk_g$ is not, since it is easily shown to be

$$u = c\{1 - (\omega_c/\omega)^2\}^{1/2} \tag{14.12}$$

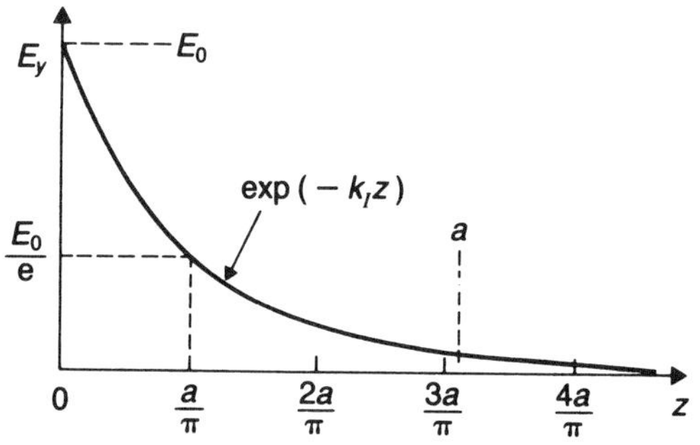

Fig. 14.4 A wave whose wavelength $\lambda_0 \gg \lambda_c$ is rapidly attenuated inside the waveguide and does not propagate.

and for all $\omega > \omega_c$, $u < c$. From (14.11) and (14.12) we get the important result

$$vu = c^2 \tag{14.13}$$

14.2.4 Magnetic field

The magnetic field for this fundamental TE mode is found from the third Maxwell equation (7.3),

$$\operatorname{curl} \boldsymbol{E} = -\partial \boldsymbol{B}/\partial t \tag{7.3}$$

Since $\boldsymbol{E} = E_y\hat{\mathbf{j}}$, the curl has only two finite components, so that

$$-\partial E_y/\partial z\,\hat{\mathbf{i}} = -\partial B_x/\partial t\,\hat{\mathbf{i}}$$

$$+\partial E_y/\partial x\,\hat{\mathbf{k}} = -\partial B_z/\partial t\,\hat{\mathbf{k}}$$

and $B_y = 0$. Substituting for E_y from (14.7) and putting $k_x = \pi/a$, we obtain

$$B_x = -\frac{k_g}{\omega} E_0 \sin\frac{\pi x}{a} \exp \mathrm{i}(\omega t - k_g z) = -\frac{k_g}{\omega} E_y \tag{14.14}$$

$$B_z = \frac{\mathrm{i}\pi}{a\omega} E_0 \cos\frac{\pi x}{a} \exp \mathrm{i}(\omega t - k_g z) \tag{14.15}$$

and

$$\frac{B_x}{B_z} = +\frac{\mathrm{i}k_g a}{\pi}\frac{\sin \pi x/a}{\cos \pi x/a} \tag{14.16}$$

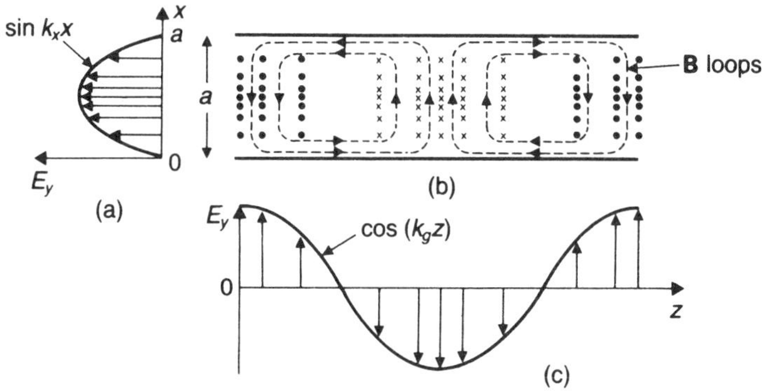

Fig. 14.5 Electric and magnetic fields of a propagating wave in the fundamental TE mode. (a) End view. (b) Central section. (c) Side view.

The electric and magnetic fields for this fundamental TE mode are drawn in Fig. 14.5 and we see that the $\boldsymbol{B}$ field forms loops which, by (14.16), are Lissajous figures for the combination of two sine waves that are of unequal amplitude and $\pi/2$ out of phase, i.e. ellipses. They are also, by (14.15), centred $\pi/2$ out of phase with the maximum of E_y.

14.2.5 Energy flow

From (14.7) and (14.14) the Poynting vector

$$\mathscr{S} = \boldsymbol{E} \times \boldsymbol{B}/\mu_0 = E_y\hat{\mathbf{j}} \times B_x\hat{\mathbf{i}}/\mu_0$$

becomes

$$\langle \mathscr{S} \rangle = k_g E_0^2 \sin^2(\pi x/a)\hat{\mathbf{k}}/(2\omega\mu_0) \qquad (14.17)$$

The total power is therefore

$$P = \frac{k_g E_0^2}{2\omega\mu_0}\int_0^a \sin^2\left(\frac{\pi x}{a}\right) b\,\mathrm{d}x$$

since $\langle \mathscr{S} \rangle$ is independent of y. Putting $v = \omega/k_g$, we find

$$P = \tfrac{1}{4}E_0^2 ab/\mu_0 v \qquad (14.18)$$

It is easy to show (exercise 1) that this power is just the electromagnetic energy per unit length multiplied by u, where u is the group (or signal) velocity given by $u = c^2/v$.

We have ignored waveguide losses by assuming perfectly conducting walls. In practice high-quality waveguides have a thin coating of silver or gold and a very small attenuation, for example 0.1 dB m^{-1} at 10 GHz, so that this loss is only significant over long distances.

14.3 WAVEGUIDE MODES

There is an infinite number of possible modes of propagation of electromagnetic waves in waveguides, but they can be classified into three basic types: (i) transverse electric (TE); (ii) transverse magnetic (TM); and (iii) transverse electric and magnetic (TEM).

14.3.1 Transverse electric modes

We have already seen that the fundamental (or dominant) TE mode in a rectangular waveguide has one component of $\boldsymbol{E}$ (E_y) only, but two components of $\boldsymbol{B}$ (B_x, B_z), producing the field distributions shown in Fig. 14.5. Higher-order TE modes have wave numbers

$$k_x = m\frac{\pi}{a}, \quad k_y = n\frac{\pi}{b} \tag{14.19}$$

for a TE_{mn} mode. The fields for such a higher-order mode must satisfy the boundary conditions and Maxwell's equations and it is easily shown using $(t, z) \equiv (\omega t - k_g z)$ that they are

$$\left.\begin{aligned} E_x &= E_{0x} \cos k_x x \sin k_y y \exp \mathrm{i}(t, z) \\ E_y &= E_{0y} \sin k_x x \cos k_y y \exp \mathrm{i}(t, z) \\ E_z &= 0 \\ B_x &= B_{0x} \sin k_x x \cos k_y y \exp \mathrm{i}(t, z) \\ B_y &= B_{0y} \cos k_x x \sin k_y y \exp \mathrm{i}(t, z) \\ B_z &= B_{0z} \cos k_x x \cos k_y y \exp \mathrm{i}(t, z) \end{aligned}\right\} \tag{14.20}$$

where the coefficients $E_{0x}, E_{0y}, B_{0x}, B_{0y}, B_{0z}$ are independent of x, y and z. The cut-off frequency is determined by substituting E_x or E_y into the wave equation (14.8), as before, giving a revised equation (14.9),

$$k_x^2 + k_y^2 + k_g^2 = \omega^2/c^2 \tag{14.21}$$

Therefore, from (14.19), the cut-off wave number is given by

$$k_c^2 = \frac{m^2\pi^2}{a^2} + \frac{n^2\pi^2}{b^2}$$

and so the cut-off frequency is

$$(v_c)_{m,n} = \frac{c}{2}\left(\frac{m^2}{a^2} + \frac{n^2}{b^2}\right)^{1/2} \tag{14.22}$$

14.3.2 Transverse magnetic modes

For these modes the magnetic fields are everywhere transverse to the direction of propagation $(0z)$, and the electric field consists of longitudinal loops. The boundary conditions require $E_z = 0$ at $x = 0$ and $y = 0$, so

$$E_z = E_{0z} \sin k_x x \sin k_y y \exp \mathrm{i}(\omega t - k_g z) \tag{14.23}$$

and the other terms can be found from Maxwell's equations and the boundary conditions, B_z being zero by definition. The resultant TM_{mn}

modes have cut-off frequencies given by (14.22), but the lowest mode is TM_{11} and there are no TM_{m0} or TM_{0n} modes, as is easily seen from (14.23). This is important in enabling one to choose a waveguide which, for a particular frequency of radiation, will then propagate *only one mode*, the TE_{10} mode (see exercise 2).

14.3.3 Transverse electric and magnetic modes

These modes have no longitudinal electric or magnetic fields and so cannot satisfy the boundary conditions for a closed, conducting waveguide. However, the 'parallel plate' waveguide of section 14.1 is bounded in one direction only and so can propagate TEM waves. The equations for propagation down a coaxial cable are normally developed from distributed capacitance and inductance formulae (see Pointon and Howarth, 'AC and DC Network Theory' in this series), but an alternative approach is to regard the wave propagating in the space between the inner diameter and outer diameter of a coaxial line as similar to that in a parallel-plate waveguide folded round into a concentric cylinder with radial distance a between the inner and outer walls (Fig. 14.6(a)). The electric field is then radial, the magnetic field azimuthal and the energy flows along $0z$ (Fig. 14.6(b)) where

$$\mathscr{S} = E_r\hat{\mathbf{r}} \times B_\phi\hat{\boldsymbol{\phi}} = \mathscr{S}_z\hat{\mathbf{z}} \tag{14.24}$$

in cylindrical polar coordinates (r, ϕ, z).

The TEM mode is the one that propagates in free space (see Fig. 9.6) and is the limiting case for a very short wavelength source (e.g. light) propagating between the parallel plates of the waveguide of Fig. 14.6(a). Then, from (14.2), as $\lambda_0 \ll 2a, \cos\theta \to 1$ and $\lambda \to \lambda_0$.

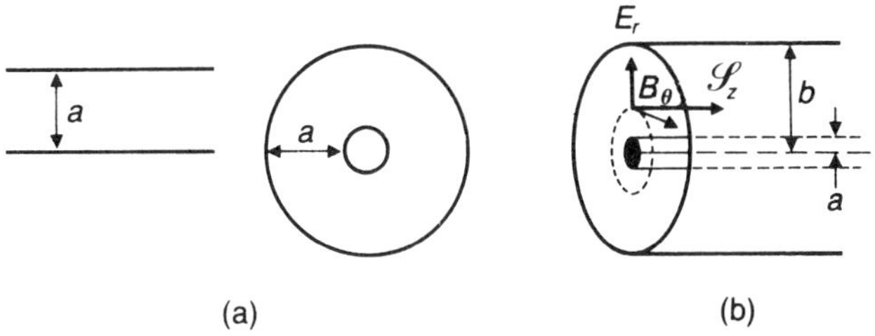

Fig. 14.6 A parallel-plate waveguide wraps round into a coaxial line. (b) The electric and magnetic fields of a TEM wave in a coaxial cable are E_r and B_θ.

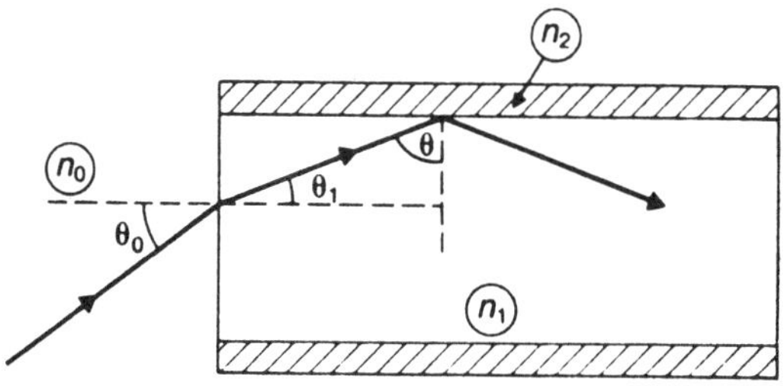

Fig. 14.7 An optical glass fibre has a core of refractive index $n_1 > n_2$, the refractive index of a thin dielectric coating, so that the light is totally internally reflected for all angles $\theta > \theta_c$, the critical angle.

The waveguide ceases to influence the propagating wave, which therefore travels at speed c, as in free space. In the same way we can see that a vacuum coaxial line with perfectly conducting walls will propagate all frequencies at speed c. In practice, the finite conductivity of the conductors in a coaxial cable produces both attenuation and dispersion, as does the dielectric in a normal cable, where the phase velocity will be $v = c/\varepsilon_r^{1/2}$ (10.25).

Recently quartz glass fibres have been developed for optical communications over terrestrial distances by reducing their attenuation to exceptionally low values, such as 1 dB km^{-1}. Such an *optical fibre* acts as a dielectric waveguide with total internal reflection at its walls achieved by applying a thin, external, dielectric coating whose refractive index is less than that of the quartz glass (Fig. 14.7). Then, from (11.52), the critical value of θ is given by

$$\sin\theta_c = n_2/n_1$$

and for all $\theta > \theta_c$ the walls will be perfectly reflecting, although an evanescent wave will penetrate the sheath. For this reason when multiple fibres form a tight bundle they are often given a second, opaque coating. Typically the fibre diameter is many optical wavelengths and so the wave that propagates is a TEM wave at phase velocity $v = c/n_1$.

14.4 CAVITIES

Closely related to travelling waves in waveguides are standing waves in resonant cavities. The simplest cavity to analyse is a hollow, rectangular one (Fig. 14.8(a)).

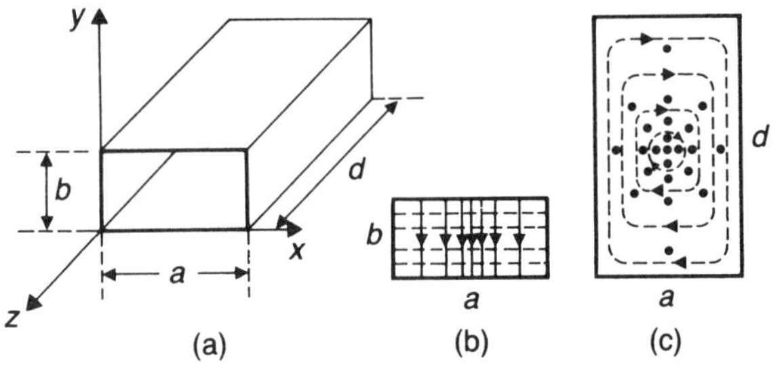

Fig. 14.8 (a) Rectangular cavity, $x=a$, $y=b$, $z=d$. (b) End view. (c) Plan view, of electric ($\longrightarrow, \cdot$) and magnetic (----) fields, for a TE_{101} mode.

14.4.1 Rectangular cavities

Since the cavity is bounded by conductors in all three directions, the resonant modes must satisfy the boundary conditions at three pairs of walls and (14.19) becomes

$$k_x = m\frac{\pi}{a}, \quad k_y = n\frac{\pi}{b}, \quad k_z = l\frac{\pi}{d} \tag{14.25}$$

for a TE_{mnl} or TM_{mnl} mode. Hence (14.20) is replaced by

$$\left.\begin{aligned} E_x &= E_{0x}\cos k_x x \sin k_y y \sin k_z z \exp i\omega t \\ E_y &= E_0 \sin k_x x \cos k_y y \sin k_z z \exp i\omega t \\ E_z &= E_0 \sin k_x x \sin k_y y \cos k_z z \exp i\omega t \end{aligned}\right\} \tag{14.26}$$

For a particular mode the *resonant frequency* will be, by analogy with (14.22),

$$v_{mnl} = \frac{c}{2}\left(\frac{m^2}{a^2} + \frac{n^2}{b^2} + \frac{l^2}{d^2}\right)^{1/2} \tag{14.27}$$

The magnetic fields can be found from the Maxwell equation $\text{curl}\,\boldsymbol{E} = -\partial \boldsymbol{B}/\partial t$, or from (14.20) by analogy. For each resonant frequency there are two possible modes (or polarizations): a TE mode and a TM mode.

There is an infinite number of these pairs of modes, but resonant cavities are normally used in one of the lower modes, for example the TE_{101} mode (Fig. 14.8(b) and (c)). Such cavities, when used at microwave frequencies, have sharp resonances that are clearly spaced. The quality factor, or Q value, of a cavity is determined from a resonance curve similar to that shown in Fig. 10.4, where in this case the 'half-width' $2\Delta\omega$ about the resonant angular frequency ω_0 is measured at the half-power points, ($-3\,\mathrm{d}B$), so that

$$Q = \omega_0/2\Delta\omega \tag{14.28}$$

Typically for a 10 GHz cavity Q is 10^4–10^5.

14.4.2 Coupling to cavities

There is a variety of ways of exciting resonant modes in cavities and, similarly, of inducing propagating waves in waveguides. Some common ones are illustrated in Fig. 14.9. A coaxial cable ending in a small antenna (wire probe) when inserted in the direction of the electric-field lines couples capacitatively to the cavity (Fig. 14.9(a)). This can be used, for example, to drive the cavity from an external oscillator. Alternatively a coaxial cable can be terminated in a small loop connected to the wall of the cavity so that the loop has its plane *normal* to the magnetic-field lines, allowing the magnetic flux to thread the loop and provide an inductive coupling (Fig. 14.9(b)). This is a particularly convenient way of both exciting a cavity and coupling out of it into a detector, for example.

When a waveguide has to be coupled to a cavity a simple iris in an appropriate plane can couple an electromagnetic wave from the waveguide into the resonant cavity. In Fig. 14.9(c) the plane is chosen so that the magnetic-field lines flow lines flow easily from one to the other, but in another case it could be the electric-field lines that have a common direction and provide the necessary coupling. Of

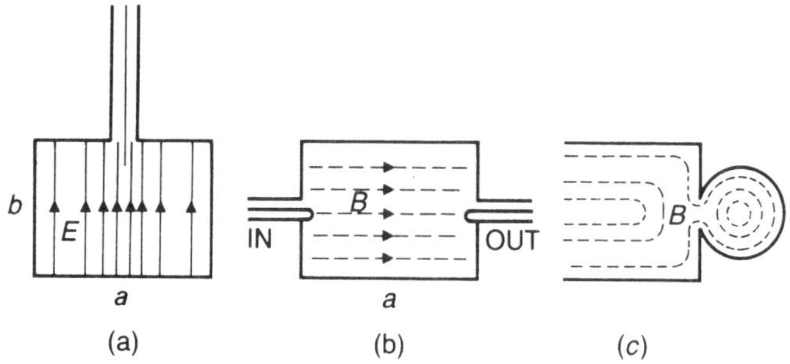

Fig. 14.9 (a) Capacitative (or electric) coupling from a coaxial cable into a cavity. (b) Inductive (or magnetic) coupling into and out of a cavity with coaxial lines. (c) Direct coupling of magnetic field in a waveguide through a small hole into a cylindrical cavity.

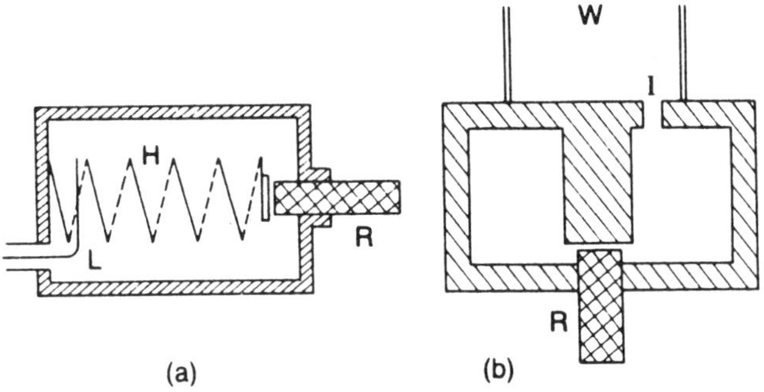

Fig. 14.10 (a) Helical resonant cavity with helix H, inductive input L and excitation of 0.1–0.5 GHz ultrasonics in the piezoelectric rod R. (b) Re-entrant cylindrical cavity with direct input through the iris I from the waveguide W and excitation of 9 GHz microsonics in the piezoelectric rod R.

course the usual requirements for impedance matching or for loose coupling have to be met.

There are many applications of resonant cavities in all branches of physics and often these are of cavities modified from the simple rectangular or cylindrical shape for particular purposes. In the author's laboratory a common use has been to generate acoustic waves at microwave frequencies. One example is a helical cavity (Fig. 14.10(a)) where the helix greatly reduces the resonant frequency of a cavity of given size and so enables small cavities to be used for the lower microwave frequencies. Here an inductive loop L couples the incoming pulses at 0.1–0.5 GHz into the cavity, which produces a high electric field at the end of a non-resonant piezoelectric rod R and so propagates a sound wave down it. A second example is a re-entrant, cylindrical cavity (Fig. 14.10(b)), which is excited with 9 GHz pulses from a wave-guide W through an iris I and similarly propagates microsonic pulses along the rod R.

14.4.3 Cavity radiation

There is an infinite number of resonant modes in a cavity and so it is important to calculate the total electromagnetic energy in a cavity at an absolute temperature T, assuming it to be thermally isolated and the radiation inside in thermal equilibrium with the cavity. To compute this total energy, we must find the number of modes between ν and $(\nu + d\nu)$ and for simplicity we will consider a cubical cavity, so that (14.27) becomes

$$\nu_{mnl} = (m^2 + n^2 + l^2)^{1/2} c/(2a) \tag{14.29}$$

To count the number of resonant modes, let ν_x, ν_y, ν_z form a set of coordinate axes and let each solution of (14.29) be a point in this frequency space. For a large volume in this space, the density of points is then $8a^3/c^3$, so that the number between spheres of radii ν and $(\nu + \mathrm{d}\nu)$ is $4\pi\nu^2\,\mathrm{d}\nu \cdot 8a^3/c^3$. However, we have seen that for each resonant frequency there is both a TE mode and a TM mode, so that the total number of modes per unit volume between ν and $(\nu + \mathrm{d}\nu)$ is

$$8\pi\nu^2\,\mathrm{d}\nu/c^3 \tag{14.30}$$

In classical statistics the principle of equipartition of energy states that each vibrational mode has an average energy of $k_B T$, where k_B is the Boltzmann constant, independent of its frequency. Therefore the energy density of cavity radiation on classical theory is

$$u(\nu, T) = 8\pi\nu^2 k_B T/c^3 \tag{14.31}$$

which is the *Rayleigh–Jeans law* for thermal radiation. It came as a shock to nineteenth-century physicists when they realized it led to an *infinite* total energy at a finite temperature, since

$$U(T) = \int_0^\infty u(\nu, T)\,\mathrm{d}\nu = \frac{8\pi k_B T}{c^3}\int_0^\infty \nu^2\,\mathrm{d}\nu \tag{14.32}$$

is infinite. This was known as the *ultraviolet catastrophe*, because it led to an infinite energy at the high-frequency end of the electromagnetic spectrum.

As every physics student knows, this led first Planck in 1901 and then Einstein in 1905 to the completely new concept of the quantization of electromagnetic energy into the now familiar *photons*, but an account of the quantum theory of electromagnetic radiation must be the subject of other texts in this series.

Further reading

The following are a few recently published texts on electromagnetism which extend the coverage of this book.

Choudhury, M.H. (1989) *Electromagnetism*, 640pp., Ellis Norwood: an advanced text for mathematicians and theoretical physicists.

Grant, I.S. and Phillips, W.R. (1990) *Electromagnetism*, Second edition, 542pp., Wiley: an undergraduate text for physicists.

Hans, H.A. and Melcher, J.R. (1989) *Electromagnetic Fields and Energy*, 742pp., Prentice Hall: an advanced text for electronic engineers.

Lorrain, P. and Corson, D.R. (1988) *Electromagnetic Fields and Waves*, Third edition, 754pp., W.H. Freeman: an undergraduate text for physicists.

Neff, H.P. (1987) *Basic Electromagnetic Fields*, 616pp., Harper and Row: mainly for electrical engineers; includes transmission lines.

The following are classical treatises on electromagnetism:

Maxwell, James Clerk (1954) *A Treatise on Electricity and Magnetism*, Reprint of the 1891 Third edition, Vol. 1, 506pp., Vol. 2, 500pp., Dover: the original theory of electromagnetism, essential reading for history of science students.

Panofsky, W.K.H. and Phillips, M. (1962) *Classical Electricity and Magnetism*, 494pp., Addition-Wesley: recognized as the authoritative advanced text for mathematicians and theoretical physicists.

The following are valuable contributions to the history of electromagnetism:

Feather, N. (1968) *Electricity and Matter*, 532pp., Edinburgh University Press: an introductory survey of the history of electricity and matter written for undergraduates.

Thomas, J.M. (1991) *Michael Faraday and the Royal Institution*, 234pp., Adam Hilger: a eulogy on the life and works on one of the greatest experimental philosophers of all time, by the present Director of the Royal Institution.

APPENDIX A

Electromagnetic quantities

Table A.1

Quantity	*Symbol*	*Units*	*Dimensions*	*Equations*
Electric current	I	A	A	SI Unit
Electric charge	q	C	AT	(4.2)
Electric dipole moment	$\boldsymbol{p}$	C m	ALT	(2.26)
Electric quadrupole moment	Q	C m^2	AL^{2}T	(13.19)
Electric field	$\boldsymbol{E}$	V m^{-1}	A^{-1}MLT^{-3}	(2.4)
Electric potential	ϕ	V	A^{-1}ML^{2}T^{-3}	(2.12)
Capacitance	C	F	A^{2}M^{-1}L^{-2}T^4	(2.19)
Electrostatic energy	U	J	ML^{2}T^{-2}	(2.36)
Electric polarization	$\boldsymbol{P}$	C m^{-2}	AL^{-2}T	$\Sigma \boldsymbol{P}$/vol
Polarizability	α	F m^2	A^{2}M^{-1}T^4	(10.3)
Electric susceptibility	χ_e	–	none	(10.1)
Relative permittivity	ε_r	–	none	(7.24), (10.32)
Electric displacement	$\boldsymbol{D}$	C m^{-2}	AL^{2}T	(2.30)
Electric charge density	ρ	C m^{-3}	AL^{-3}T	q/vol
Linear charge density	λ	C m^{-1}	AL^{-1}T	q/length

(*Contd.*)

Table A.1 (*Contd.*)

Quantity	*Symbol*	*Units*	*Dimensions*	*Equations*
Surface charge density	σ	$\mathrm{C\,m^{-2}}$	$\mathrm{AL^{-2}T}$	q/area
Electric current density	$\boldsymbol{j}$	$\mathrm{A\,m^{-2}}$	$\mathrm{AL^{-2}}$	q/area/s
Surface current density	$\boldsymbol{i}$	$\mathrm{A\,m^{-1}}$	$\mathrm{AL^{-1}}$	(7.17)
Electrical conductivity	γ	$\mathrm{S\,m^{-1}}$	$\mathrm{A^2M^{-1}L^{-3}T^3}$	(12.1)
Electrical resistance	R	Ω	$\mathrm{A^{-2}ML^2T^{-3}}$	(4.7)
Electromotive force	$\mathscr{E}$	V	$\mathrm{A^{-1}ML^2T^{-3}}$	(5.1)
Magnetic field	$\boldsymbol{B}$	T	$\mathrm{A^{-1}MT^{-2}}$	(4.8)
Magnetic dipole moment	$\boldsymbol{m}$	$\mathrm{A\,m^2}$	$\mathrm{AL^2}$	(4.12)
Magnetic flux	Φ	Wb	$\mathrm{A^{-1}ML^2T^{-2}}$	(4.28)
Inductance, mutual	M	H	$\mathrm{A^{-2}MLT^{-2}}$	(5.11)
Inductance, self	L	H	$\mathrm{A^{-2}MLT^{-2}}$	(5.14)
Magnetization	$\boldsymbol{M}$	$\mathrm{A\,m^{-1}}$	$\mathrm{AL^{-1}}$	$\Sigma\boldsymbol{m}$/vol
Magnetizing field	$\boldsymbol{H}$	$\mathrm{A\,m^{-1}}$	$\mathrm{AL^{-1}}$	(6.7)
Magnetic vector potential	$\boldsymbol{A}$	$\mathrm{Wb\,m^{-1}}$	$\mathrm{A^{-1}MLT^{-2}}$	(8.41)
Magnetic susceptibility	χ_{m}	–	none	$\boldsymbol{M}/\boldsymbol{H}$
Relative permeability	μ_{r}	–	none	(7.22)
Magnetostatic energy	U	J	$\mathrm{ML^2T^{-2}}$	(6.15)
Larmor frequency	ω_{L}	$\mathrm{s^{-1}}$	$\mathrm{T^{-1}}$	(6.20)
Magnetomotive force	$\mathscr{F}_{\mathrm{m}}$	A	A	(6.29)
Magnetic reluctance	$\mathscr{R}$	$\mathrm{A\,Wb^{-1}}$	$\mathrm{A^2M^{-1}L^{-2}T^2}$	(6.29)
Electromagnetic energy density	u	$\mathrm{J\,m^{-3}}$	$\mathrm{ML^{-1}T^{-2}}$	(9.30), (11.40)
Poynting vector	$\mathscr{S}$	$\mathrm{W\,m^{-2}}$	$\mathrm{MT^{-3}}$	(9.29), (11.39)
Wave impedance	Z	Ω	$\mathrm{A^{-2}ML^2T^{-3}}$	(9.36)
Refractive index	n	–	none	(10.26), (10.29)

Table A.1 (*Contd.*)

Quantity	*Symbol*	*Units*	*Dimensions*	*Equations*
Wave number	k	m^{-1}	L^{-1}	(9.10), (12.10)
Absorption coefficient	β	m^{-1}	L^{-1}	(10.31)
Skin depth	δ	m	L	(12.13)
Reflectance	R	–	none	(11.45)
Transmittance	T	–	none	(11.46)
Plasma frequency	ω_p	s^{-1}	T^{-1}	(12.35)
Radiation pressure	p_r	Pa ($J\,m^{-3}$)	$ML^{-1}T^{-2}$	(12.27)
Radiated power	P	W	ML^2T^3	(13.12)

APPENDIX B

Gaussian units

The Système International d'Unités (SI) used in this book is that adopted by the General Conferences of Weights and Measures (CGPM) and endorsed by the International Organisation for Standardization (ISO) for use by engineers and scientists throughout the world. It is based on six fundamental units: metre (m), kilogramme (kg), second (s), ampère (A), kelvin (K) and candela (cd). Such a system is, of course, arbitrary and its chief merit is that it is agreed internationally. For convenience in theoretical physics two other systems of units are often chosen: natural and Gaussian units. In the system of natural units the universal constants (h, k_B, c) are chosen to be dimensionless and of unit size, which is useful in the theory of elementary particles. In the Gaussian system the older metric units of centimetre, gram and second (c.g.s.) are retained with an electrostatic unit (e.s.u.) for electric charge and an electromagnetic unit (e.m.u.) for electric current. The ratio of current (in e.s.u.) to current (in e.m.u.) has the dimensions of a velocity and is the velocity of light *in vacuo*, c (in c.g.s. units). The net results are that ε_0 and μ_0 are dimensionless and of unit size in this system, leading to the replacement of $(\varepsilon_0\mu_0)^{-1/2}$ by c, and that the absence of $1/4\pi$ in Coulomb's law of force leads to the presence of 4π in terms involving charges and currents. The main advantages of this system are that $\boldsymbol{E}$ and $\boldsymbol{B}$ have the same dimensions and are of equal magnitude for electromagnetic waves in free space. However, in media $\boldsymbol{D} = \varepsilon \boldsymbol{E}$ and $\boldsymbol{B} = \mu$H and some of this simplicity is lost.

Some of the important equations in electromagnetism are given in Table B.1 in Gaussian units, the equation number being that for SI units in the text. Conversions of some Gaussian units to SI units are given in Table B.2, assuming $c = 3 \times 10^8\,\mathrm{m\,s^{-1}}$.

Table B.1 *Electromagnetic equations in Gaussian units*

Maxwell I	$\text{div}\,\boldsymbol{E} = 4\pi\rho/\varepsilon$	(7.1); $\text{div}\,\boldsymbol{D} = 4\pi\rho$	(7.12)
Maxwell II	$\text{div}\,\boldsymbol{B} = 0$	(7.2)	
Maxwell III	$\text{curl}\,\boldsymbol{E} = -\frac{1}{c}\frac{\partial \boldsymbol{B}}{\partial t}$	(7.3)	
Maxwell IV	$\text{curl}\,\boldsymbol{B} = \frac{4\pi}{c}\boldsymbol{j} + \frac{1}{c^2}\frac{\partial \boldsymbol{E}}{\partial t}$	(7.4); $\text{curl}\,\boldsymbol{H} = \frac{4\pi}{c}\boldsymbol{j}_\mathrm{f} + \frac{1}{c}\frac{\partial \boldsymbol{D}}{\partial t}$	(7.22)
Lorentz force		$\boldsymbol{F} = q\left(\boldsymbol{E} + \frac{\boldsymbol{u}\times\boldsymbol{B}}{c}\right)$	(4.9)
Electric displacement		$\boldsymbol{D} = \boldsymbol{E} + 4\pi\boldsymbol{P}$	(2.30)
Magnetizing field		$\boldsymbol{H} = \boldsymbol{B} - 4\pi\boldsymbol{M}$	(6.7)
Electric susceptibility		$\chi_\mathrm{e} = \frac{1}{4\pi}(\varepsilon - 1)$	(10.1)
Magnetic susceptibility		$\chi_\mathrm{m} = \frac{1}{4\pi}(\mu - 1)$	(6.8)
Energy density		$u = (1/8\pi)(\boldsymbol{E}.\boldsymbol{D} + \boldsymbol{B}.\boldsymbol{H})$	(11.40)

Table B.2 *Conversion of Gaussian units to SI units*

Electric charge	q	3×10^9 e.s.u. = 1 coulomb
Electric current	I	1 e.m.u. = 10 ampere
Electric potential	ϕ	1 stat volt = 300 volt
Electric field	$\boldsymbol{E}$	3×10^4 stat volt cm^{-1} = 1 volt metre^{-1}
Electric displacement	$\boldsymbol{D}$	$12\pi\times10^5$ e.s.u. cm^{-2} = 1 coulomb metre^{-2}
Energy density	u	10^{13} erg cm^{-3} = 1 joule metre^{-3}
Radiated power	P	10^7 erg second^{-1} = 1 watt
Resistance	R	1 stat ohm = 9×10^{11} ohm
Capacitance	C	9×10^{11} cm = 1 farad
Inductance	L	10^9 e.m.u. = 1 henry
Magnetic field	$\boldsymbol{B}$	10^4 gauss = 1 tesla
Magnetizing field	$\boldsymbol{H}$	$4\pi\times10^{-3}$ oersted = 1 ampere metre^{-1}

APPENDIX C
Physical constants

Table C.1

Constant	*Symbol*	*Value*
Electric constant	$\varepsilon_0 = 1/(\mu_0 c^2)$	$8.85 \times 10^{-12}\,\mathrm{F\,m^{-1}}$
Magnetic constant	μ_0	$4\pi \times 10^{-7}\,\mathrm{H\,m^{-1}}$
Speed of light	c	$3.00 \times 10^{8}\,\mathrm{m\,s^{-1}}$
Electronic charge	e	$1.60 \times 10^{-19}\,\mathrm{C}$
Rest mass of electron	m_e	$9.11 \times 10^{-31}\,\mathrm{kg}$
Rest mass of proton	m_p	$1.67 \times 10^{-27}\,\mathrm{kg}$
Planck constant	h	$6.63 \times 10^{-34}\,\mathrm{J\,s}$
	$\hbar = h/2\pi$	$1.05 \times 10^{-34}\,\mathrm{J\,s}$
Boltzmann constant	k_B	$1.38 \times 10^{-23}\,\mathrm{J/K^{-1}}$
Avogadro constant	N_A	$6.02 \times 10^{23}\,\mathrm{mol^{-1}}$
Gravitation constant	G	$6.67 \times 10^{-11}\,\mathrm{N\,m^2\,kg^{-2}}$
Bohr radius	$a_0 = \dfrac{4\pi\varepsilon_0 \hbar^2}{m_e^2}$	$5.29 \times 10^{-11}\,\mathrm{m}$
Bohr magneton	$\mu_B = \dfrac{e\hbar}{2m_e}$	$9.27 \times 10^{-24}\,\mathrm{J\,T^{-1}}$
Electron volt	eV	$1.60 \times 10^{-19}\,\mathrm{J}$
Molar volume at S.T.P.	V_m	$2.24 \times 10^{-2}\,\mathrm{m^3\,mol^{-1}}$
Acceleration due to gravity	g	$9.81\,\mathrm{m\,s^{-2}}$

APPENDIX D

Electromagnetic spectrum

	Frequency ν Hz		*Wavelength* λ m	SPECTRAL REGIONS		*Energy* $h\nu$ J	*Temperature* $h\nu/k_B$
	⋮		⋮			⋮	
	10^{24}	1 fm	10^{-15}	Cosmic ray	1 GeV	10^{-10}	10^{13} K
	10^{21}	1 pm	10^{-12}	Gamma ray	1 MeV	10^{-13}	10^{10} K
1 EHz	10^{18}	1 nm	10^{-9}	X-ray	1 keV	10^{-16}	10^{7} K
1 PHz	10^{15}	1 μm	10^{-6}	Ultraviolet Visible Infrared	1 eV	10^{-19}	10^{4} K
							Room
1 THz	10^{12}	1 mm	10^{-3}	Millimetre wave Microwave	1 meV	10^{-22}	10 K
1 GHz	10^{9}	1 m	1	Television FM radio	1 μeV	10^{-25}	10 mK
1 MHz	10^{6}	1 km	10^{3}	AM radio	1 neV	10^{-28}	10 μK
1 kHz	10^{3}	1 Mm	10^{6}	Very low frequency	1 peV	10^{-31}	10 nK

Wave Properties | *Particle Properties*

$$[1\ \text{Hz} \equiv 3.00 \times 10^{8}\ \text{m} \equiv 4.14 \times 10^{-15}\ \text{eV} \equiv 4.80 \times 10^{-11}\ \text{K}]$$

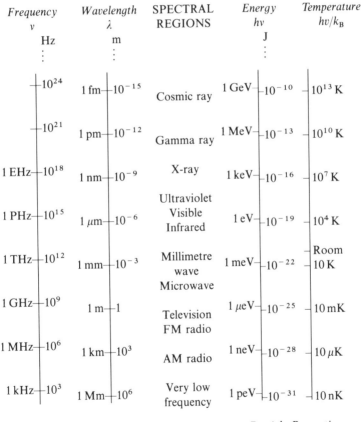

APPENDIX E

Vector calculus

Scalar and vector fields may be operated on by the differential operators grad, div, curl and ∇^2. Expressions for the results of these operators on general vectors $\boldsymbol{A}, \boldsymbol{B}$ and a scalar Ω are given below for Cartesian coordinates (x, y, z), spherical polar coordinates (r, θ, ψ), Fig. 2.5, and cylindrical polar coordinates (r, ϕ, z), Fig. 3.2.

Identities

$$\operatorname{div}(\Omega \boldsymbol{A}) = \Omega \operatorname{div} \boldsymbol{A} + \boldsymbol{A} \cdot \operatorname{grad} \Omega \tag{E.1}$$

$$\operatorname{div}(\boldsymbol{A} \times \boldsymbol{B}) = \boldsymbol{B} \cdot \operatorname{curl} \boldsymbol{A} - \boldsymbol{A} \cdot \operatorname{curl} \boldsymbol{B} \tag{E.2}$$

$$\operatorname{curl}(\Omega \boldsymbol{A}) = \Omega \operatorname{curl} \boldsymbol{A} + \operatorname{grad} \Omega \times \boldsymbol{A} \tag{E.3}$$

$$\operatorname{curl} \operatorname{grad} \Omega = 0 \tag{E.4}$$

$$\operatorname{div} \operatorname{curl} \boldsymbol{A} = 0 \tag{E.5}$$

$$\operatorname{curl} \operatorname{curl} \boldsymbol{A} = \operatorname{grad} \operatorname{div} \boldsymbol{A} - \nabla^2 \boldsymbol{A} \tag{E.6}$$

Cartesian differential operators

$$\operatorname{grad} \Omega = \nabla \Omega = \frac{\partial \Omega}{\partial x} \hat{\mathbf{i}} + \frac{\partial \Omega}{\partial y} \hat{\mathbf{j}} + \frac{\partial \Omega}{\partial z} \hat{\mathbf{k}} \tag{E.7}$$

$$\operatorname{div} \boldsymbol{A} = \nabla \cdot \boldsymbol{A} = \frac{\partial A_x}{\partial x} + \frac{\partial A_y}{\partial y} + \frac{\partial A_z}{\partial z} \tag{E.8}$$

$$\operatorname{curl} \boldsymbol{A} = \nabla \times \boldsymbol{A} = \begin{vmatrix} \hat{\mathbf{i}} & \hat{\mathbf{j}} & \hat{\mathbf{k}} \\ \partial/\partial x & \partial/\partial y & \partial/\partial z \\ A_x & A_y & A_z \end{vmatrix}. \tag{E.9}$$

$$\operatorname{div}(\operatorname{grad}) = \nabla^2 = \frac{\partial^2}{\partial x^2} + \frac{\partial^2}{\partial y^2} + \frac{\partial^2}{\partial z^2} \tag{E.10}$$

Spherical polar differential operators

$$\operatorname{grad} \Omega = \frac{\partial \Omega}{\partial r}\hat{\mathbf{r}} + \frac{1}{r}\frac{\partial \Omega}{\partial \theta}\hat{\boldsymbol{\theta}} + \frac{1}{r\sin\theta}\frac{\partial \Omega}{\partial \psi}\hat{\boldsymbol{\psi}} \tag{E.11}$$

$$\operatorname{div} \boldsymbol{A} = \frac{1}{r^2}\frac{\partial}{\partial r}(r^2 A_r) + \frac{1}{r\sin\theta}\frac{\partial}{\partial \theta}(A_0 \sin\theta) + \frac{1}{r\sin\theta}\frac{\partial A_\psi}{\partial \psi} \tag{E.12}$$

$$\operatorname{curl} \boldsymbol{A} = \frac{1}{r^2\sin\theta}\begin{vmatrix} \hat{\mathbf{r}} & r\hat{\boldsymbol{\theta}} & r\sin\theta\,\hat{\boldsymbol{\psi}} \\ \partial/\partial r & \partial/\partial\theta & \partial/\partial\psi \\ A_r & rA_\theta & r\sin\theta A_\psi \end{vmatrix} \tag{E.13}$$

$$\nabla^2 = \frac{1}{r^2}\frac{\partial}{\partial r}\left(r^2\frac{\partial}{\partial r}\right) + \frac{1}{r^2\sin\theta}\frac{\partial}{\partial\theta}\left(\sin\theta\frac{\partial}{\partial\theta}\right) + \frac{1}{r^2\sin^2\theta}\frac{\partial^2}{\partial\psi^2} \tag{E.14}$$

Cylindrical polar differential operators

$$\operatorname{grad} \Omega = \frac{\partial \Omega}{\partial r}\hat{\mathbf{r}} + \frac{\partial \Omega}{r\partial \phi}\hat{\boldsymbol{\phi}} + \frac{\partial \Omega}{\partial z}\hat{\mathbf{z}} \tag{E.15}$$

$$\operatorname{div} \boldsymbol{A} = \frac{1}{r}\frac{\partial}{\partial r}(rA_r) + \frac{\partial A_\phi}{r\partial\phi} + \frac{\partial A_z}{\partial z} \tag{E.16}$$

$$\operatorname{curl} \boldsymbol{A} = \frac{1}{r}\begin{vmatrix} \hat{\mathbf{r}} & \mathbf{r}\hat{\boldsymbol{\phi}} & \hat{\mathbf{z}} \\ \partial'/\partial r & \partial/\partial\phi & \partial/\partial z \\ A_r & rA_\phi & A_z \end{vmatrix} \tag{E.17}$$

$$\nabla^2 = \frac{1}{r}\frac{\partial}{\partial r}\left(r\frac{\partial}{\partial r}\right) + \frac{1}{r^2}\frac{\partial^2}{\partial\phi^2} + \frac{\partial^2}{\partial z^2} \tag{E.18}$$

Theorems

For a smoothly varying vector field $\boldsymbol{A}$.

Gauss's divergence theorem

$$\int_S \boldsymbol{A}\cdot\mathbf{d}\boldsymbol{S} = \int_V \operatorname{div} \boldsymbol{A}\,\mathrm{d}\tau \tag{E.19}$$

where the surface S encloses the volume V, $\mathbf{d}\boldsymbol{S} = \hat{\mathbf{n}}\,\mathrm{d}S$ is a vector of magnitude $\mathrm{d}S$ along the outward normal $\hat{\mathbf{n}}$ to the surface $\mathrm{d}S$ and $\mathrm{d}\tau$ is an element of the volume V.

Stokes's theorem

$$\oint_C \boldsymbol{A}\cdot\mathbf{d}\boldsymbol{s} = \int_S (\mathrm{curl}\,\boldsymbol{A})\cdot\mathbf{d}\boldsymbol{S} \qquad \text{(E.20)}$$

where the closed loop C bounds the surface S and $\mathbf{d}\boldsymbol{s}$ is a vector element of the loop C.

APPENDIX F

Lorentz transformations

The transformation of physical quantities from an inertial frame S (the laboratory frame) to a frame S' moving with respect to S at a speed u in the positive x direction is given in Cartesian coordinates, where $\beta = u/c$ and $\gamma = (1-\beta^2)^{-1/2}$. For the inverse transforms replace u by $-u$. (See also Chapter 8 and Fig. 8.1.)

Coordinates

$$x' = \gamma(x - ut), \quad y' = y, \quad z' = z, \quad t' = \gamma\left\{t - \left(\frac{\beta}{c}\right)x\right\} \tag{F.1}$$

Velocities

$$v'_x = \frac{v_x - u}{\{1-(\beta/c)v_x\}}, \quad v'_y = \frac{v_y}{\gamma\{1-(\beta/c)v_x\}}, \quad v'_z = \frac{v_z}{\gamma\{1-(\beta/c)v_x\}} \tag{F.2}$$

Components of a force

$$F'_x = F_x - \frac{(\beta/c)(v_y F_y + v_z F_z)}{\{1-(\beta/c)v_x\}} \tag{F.3}$$

$$F'_y = \frac{F_y}{\gamma\{1-(\beta/c)v_x\}} \tag{F.4}$$

$$F'_z = \frac{F_z}{\gamma\{1-(\beta/c)v_x\}} \tag{F.5}$$

Electric field

$$E'_x = E_x, \quad E'_y = \gamma(E_y - \beta c B_z), \quad E'_z = \gamma(E_z + \beta c B_y) \tag{F.6}$$

Magnetic field

$$B'_x = B_x, \quad B'_y = \gamma\{B_y + (\beta/c)E_z\}, \quad B'_z = \gamma\{B_z - (\beta/c)E_y\} \tag{F.7}$$

APPENDIX G

Exercises

(L) indicates University of London question.

Chapter 2

1 Compare the magnitudes of the electrostatic and gravitational forces between two alphas particles (He^{2+}).

2 A charge of 111 pC is uniformly distributed throughout the volume of an isolated sphere of diameter 40 cm. Calculate the electric field at the following distances from the centre of the sphere: (a) zero; (b) 10 cm; (c) 20 cm; (d) 50 cm. (L)

3 How much work is done on an electron when it is moved from $(3, 2, -1)$ to $(2, 1, -4)$ in an electric field given by $\boldsymbol{E} = (3\mathbf{i} - 4\mathbf{j} + 2\mathbf{k})\ \text{V cm}^{-1}$? (Distance in metres). (L)

4 A charge of 3 μC is uniformly distributed along a thin rod of length 40 cm. Find the electric field at a point 20 cm from the rod on its perpendicular bisector.

5 Show from first principles that the dimensions of capacitance are $A^2M^{-1}L^{-2}T^4$ on the MKSA (SI) system and find the dimensions of the electric constant, ε_0.

6 An electron moves from rest through a displacement $(10^{-3}\hat{\mathbf{i}} + 10^{-3}\hat{\mathbf{j}} + 10^{-3}\hat{\mathbf{k}})$ metres within an electric field $(10^5\hat{\mathbf{i}} + 10^5\hat{\mathbf{j}} + 10^5\hat{\mathbf{k}})\ \text{V m}^{-1}$. Find (a) the kinetic energy gained by the electron, and (b) its final velocity.

7 Prove that the capacitance of an isolated sphere of radius a is $4\pi\varepsilon_0 a$.

8 Eight identical spherical drops of mercury are each charged to 10 V above earth (ground) potential and then allowed to coalesce into a single spherical drop. What is the potential of

the large drop? How has the electrostatic energy of the system changed? (*L*)

9 A spherical electrode is required to carry a charge of 33 nC. Estimate its minimum radius if the breakdown field strength of the surrounding air is $3 \times 10^6\,\mathrm{V\,m^{-1}}$.

10 Two isolated metal spheres of radii 40 and 90 mm are charged to 0.9 and 2.0 kV, respectively. They are then connected by a fine wire. Explain what happens to (a) the electric charges, (b) the electric potentials, and (c) the electrical energy stored in the system. (d) Hence find the stored energy lost as a result of the connection.

11 A 12 pF parallel-plate air capacitor is charged by connecting it to a 100 V battery. How much work must be done to double the separation of the plates of the capacitor (a) with the battery connected, and (b) with it disconnected and the capacitor fully charged? Explain why the answers differ.

12 A large parallel plate capacitor is completely filled by a 2 mm thick slab of a dielectric ($\varepsilon_r = 6$) and a 1 mm thick slab of another dielectric ($\varepsilon_r = 2$). The plates are connected to a 1 kV battery, the plate next to the thick layer being positive and the other earthed (grounded). Calculate (a) the surface charge density on the plates, and (b) the potential at the dielectric interface. (*L*)

13 A long coaxial cable consists of an inner wire radius a inside a metal tube of inside radius b, the space in between being completely filled with a dielectric of relative permittivity ε_r. Show that its capacitance is $(2\pi\varepsilon_r\varepsilon_0)/(ln b/a)\,\mathrm{F\,m^{-1}}$. (Hint: Use Gauss's law for the electric displacement flux.)

14 A parallel plate capacitor consists of rectangular plates of length a and width b spaced d apart connected to a battery of voltage V. Arectangular slab of dielectric of relative permittivity ε_r that would just fill the space between the capacitor plates is inserted part way between them, so that it resembles a partly opened matchbox. Show that the force pulling it into the plates is $(\varepsilon_r - 1)(\varepsilon_0 b V^2)/(2d)$ and explain why it is independent of a.

Chapter 3

1 An electric dipole of moment $\boldsymbol{p} = q\boldsymbol{a}$ is placed with its centre at the origin and its axis along $\theta = 0$, where (r, θ, ψ) are spherical polar coordinates. Show that the potential $\phi(r, \theta, \psi)$, for $r \gg a$,

is independent of ψ and given by $(\boldsymbol{p}\cdot\boldsymbol{r})/(4\pi\varepsilon_0 r^3)$. Hence show that the electric field at a point (r, θ) has amplitude

$$\frac{p}{4\pi\varepsilon_0 r^3}(1 + 3\cos^2\theta)^{1/2} \qquad (L)$$

2 Using the expression $\phi = p\cos\theta/4\pi\varepsilon_0 r^2$, plot the equipotentials and field lines for a dipole and compare them with those for a point charge.

3 A pair of electric dipoles are placed end-on in a straight line so that their negative charges coincide and they form a linear quadrupole of length $2a$ with charges $+q$ at each end and $-2q$ in the centre. Given that the electric field on the axis of a dipole at a distance $r \gg a$ is $(2qa)/(4\pi\varepsilon_0 r^2)$, find the electric field on the axis of the quadrupole at a distance $r(r \gg a)$ from its centre.

4 Show that a dipole of moment $\boldsymbol{p}$ placed in a uniform electric field $\boldsymbol{E}$ acquires a potential energy $-\boldsymbol{p}\cdot\boldsymbol{E}$ and that a couple $\boldsymbol{p} \times \boldsymbol{E}$ acts to align it with the field.

5 A conducting sphere of radius R is placed in a uniform electric field $\boldsymbol{E}_0$. Show that the potential at a point outside the sphere and at (r, θ) from its centre is

$$-E_0 r\cos\theta\{1 - (R^3)/(r^3)\}$$

and hence find the components E_r and E_θ of the electric field at that point.

6 An isotropic dielectric sphere of radius R and relative permittivity ε_1, is placed in a medium of relative permittivity ε_2 containing a uniform electric field $\boldsymbol{E}_0$. Show that the potentials at points (r, θ) from the centre of the sphere are

(a) $$-3\varepsilon_2 E_0 r\cos\theta/(\varepsilon_1 + 2\varepsilon_2)$$

inside the sphere and

(b) $$-E_0 r\cos\theta\left\{1 - \left(\frac{\varepsilon_1 - \varepsilon_2}{\varepsilon_1 + 2\varepsilon_2}\right)\frac{R^3}{r^3}\right\}$$

outside it. Hence draw the lines of electric displacement $\boldsymbol{D}$ inside and outside the sphere when $\varepsilon_1 < \varepsilon_2$ and $\varepsilon_1 > \varepsilon_2$. (Hint: Assume that the potentials are of the same form as that for the conducting sphere and apply the boundary conditions to obtain the coefficients in each case.)

7 A long cylindrical conductor of radius R is earthéd and placed in a uniform electric field $\boldsymbol{E}_0$ with its axis normal to $\boldsymbol{E}_0$. Show

that the potential at a point (r, ϕ, z) is

$$V = E_0 r \cos\phi \{1 - (a^2)/(r^2)\}$$

(Hint: Show that this solution satisfies Laplace's equation and the boundary conditions.)

8 Show that a positive charge placed at (4, 2) between conducting places $y = 0$ and $x = y$ produces the same electric field between the planes as a system of four positive and four negative charges and find the positions of these charges. Will this method work for any angle between the planes?

9 Use the method of images to show that an uncharged, insulated conducting sphere of radius R is attracted to a positive charge q_1 at a distance r from its centre by the force

$$\frac{q_1^2}{4\pi\varepsilon_0}\left[\frac{R}{r^3} - \frac{R}{r\{r - (R^2)/(r)\}^2}\right]$$

(Hint: Find the image charge for an earthed (grounded) sphere first.)

10 A dipole of moment $\boldsymbol{p}$ is placed at a distance r from the centre of an earthed (grounded) conducting sphere of radius $R < r$. The axis of the dipole is in the direction from the centre of the sphere to the centre of the dipole. Prove that the image system for the dipole is a point charge pR/r^2 plus a dipole of strength pR^3/r^3 and find their positions.

Chapter 4

1 The current density in a long conductor of circular cross-section and radius R varies with radius as $j = j_0 r^2$. Calculate the current flowing in the conductor.

2 An earthing (grounding) plate is formed from a hemispherical spinning of copper sheet placed in the earth with its rim at the earth's surface. Calculate its earthing resistance, given that its diameter is 0.5 m and the conductivity of the earth is 0.01 S m^{-1}.

3 Calculate the magnitude and direction of the magnetic field at the centre of a short solenoid of radius 3 cm and length 8 cm carrying a current of 10 A and having 8 turns per cm. (*L*)

4 Show that the magnetic field at the ends and on the axis of a long solenoid is half that its centre.

5 A Helmholtz pair consists of two identical circular coils of radius a carrying the same current I in the same direction mounted

coaxially an optimum distance b apart to give the maximum volume of uniform magnetic field between them. Show that this is achieved when $b = a$ and find the uniform field. (Hint: If the axis of the coils is the x-axis, show that $\partial B/\partial x$ and $\partial^2 B/\partial x^2$ are zero for B on this axis.)

6 An electron is moving at $2 \times 10^6\ \mathrm{m\,s^{-1}}$ round a circular orbit of radius 5×10^{-11} m. What is the magnetic field at the centre of the orbit?

7 Calculate the force per unit length on each of a pair of parallel wires 5 cm apart carrying the same current of 500 A. What happens when one of the currents is reversed? *(L)*

8 Calculate the magnetic field at the centre of a square wire loop of sides 10 cm in length carrying a current of 10 A. (Hint: Consider each side separately.)

9 A steady current I flows in one direction in the solid inner conductor, radius a, of a coaxial cable and in the opposite direction in the outer conductor, of inner radius b and outer radius c. Find the magnetic field at a distance r from the axis in each of the following regions: (a) $r \leqslant a$; (b) $a \leqslant r \leqslant b$; (c) $b \leqslant r \leqslant c$; (d) $r \geqslant c$.

10 Find the magnetic dipole moment μ_B of an electron moving in an orbit with a Bohr radius a_0.

Chapter 5

1 A current loop of area S in the xy plane is placed in a uniform magnetic field $B_z = B_0 \sin \omega t$. Find the e.m.f. induced in the coil when: (a) it is fixed; (b) it rotates at an angular frequency ω.

2 An aircraft with a wingspan of 50 m flies horizontally at $300\ \mathrm{m\,s^{-1}}$ through a vertical magnetic field of 20 μT. What is the potential difference between the wing tips?

3 Estimate the kinetic energy in MeV of the electrons in a betatron when they are travelling along an orbit of radius 0.25 m in a magnetic field of 0.5 T. (Hint: High-speed electrons travel at nearly the speed of light.)

4 A thin metal disc of 5 cm radius is mounted on an axle of 1 cm radius and rotated at 2400 revolutions per minute in a uniform magnetic field of 0.4 T parallel to the axle. Brushes, connected in parallel, make contact with many points of the rim of the disc, and another set of brushes make similar contacts with the axle. The resistance measured between the sets of brushes due to the

disc is 100 μΩ. Calculate the magnitude and direction of the current flowing in the disc. (*L*)

5 A solenoid of 1000 turns is 2 cm in diameter and 10 cm long. Estimate the energy stored in it when it is carrying 100 A. (*L*)

6 Calculate the mutual inductance between a very long solenoid with 200 turns per metre and a circular coil consisting of 50 turns of mean area 20 cm^2 placed coaxially inside the solenoid. (*L*)

7 A large solenoid of radius 5 cm is wound uniformly with 1000 turns per metre and has a secondary winding of 500 turns closely wound over its centre portion. If a sinusoidal current of 2 A r.m.s. and frequency 50 Hz is passed through the solenoid, what is the e.m.f. induced in the secondary?

8 An ignition coil consisting of 16 000 turns wound closely on a solenoid of 400 turns and length 10 cm has a radius of 3 cm. If the primary current of 3 A is broken 10 000 times each second by the car's motion, what is the voltage produced across the spark plugs?

9 A long transmission line consists of two thin, parallel metal strips each 1 cm wide facing each other across a 5 cm gap. What is the inductance per metre of the line? (Hint: Consider the strips to be segments of an infinite circular, coaxial line.)

Chapter 6

1 A long paramagnetic circular rod of diameter 6 mm is suspended vertically from a balance so that one end is between the poles of an electromagnet providing a horizontal magnetic field of 1.5 T and the other end is in a negligible field. If the apparent increase in mass of the rod is 5 g, what is the magnetic susceptibility of the material? (Hint: Use the principle of virtual work to relate force to energy.)

2 Calculate the total magnetic energy in the earth's external field by assuming it is due to a dipole at the centre of the earth producing a magnetic field of 100 μT at the equator. (Assume the radius of the earth is 6400 km and integrate the dipolar field external to the earth.)

3 State whether the following statements are true or false in the MKSA system. (a) $\boldsymbol{B}$ and $\boldsymbol{H}$ are the same physical quantity but are measured in different units. (b) In free space $\boldsymbol{B} = \mu_0 \boldsymbol{H}$ means that $\boldsymbol{B}$ measures the response of a vacuum to the applied field $\boldsymbol{H}$. (c) In matter $\boldsymbol{H}$ is the volume average of the microscopic value of $\boldsymbol{B}$, multiplied by a factor independent of the material. (d) In

matter if $\boldsymbol{H}, \boldsymbol{B}$ and $\boldsymbol{M}$ are smoothed to remove irregularities on an atomic scale, then $\boldsymbol{B}$ is the sum of $\boldsymbol{H}$ and $\boldsymbol{M}$, multiplied by a factor independent of the material. (e) The sources of $\boldsymbol{H}$ are magnetic poles and of $\boldsymbol{B}$ are electric currents.

4 An intense magnetic field pulse is generated by discharging rapidly a bank of 1000 5 μF capacitors, wired in parallel, through a short, strong coil of inductance 10 mH and volume 10 cm^3. If the capacitors are charged to 1.5 kV before discharge, calculate (a) the maximum magnetic field produced, (b) the maximum current, and (c) the time to reach the maximum field. (Hint: Assume a sinusoidal discharge.)

5 An electromagnet consists of a soft iron ring of mean radius 8 cm with an air gap of 3 mm. It is wound with 100 turns carrying a current of 3 A. Given that the B–H curve for soft iron has the following characteristics, determine the magnetic field in the gap:

B(T)	0.2	0.4	0.6	0.8	1.0
H(A m^{-1})	100	170	220	310	500

Chapter 8

1 Use the Lorentz transformations for the components of a force, of a velocity and of the coordinates, given in Appendix F, to show that the Coulomb force $\boldsymbol{F}'$ on a moving charge in a moving frame, (8.14), transforms into the Lorentz force $\boldsymbol{F}$, (8.15), in the laboratory frame.

2 A narrow beam of electrons of energy 7 GeV is circulating in a large storage ring. What are the maximum instantaneous values of the electric and magnetic fields due to each electron at a distance of 5 mm from the beam and in which direction is each vector relative to the beam?

3 Show that the vector potential at a point P, distance $\boldsymbol{r}$ from the centre of a thin, circular wire carrying a persistent current I, is

$$\boldsymbol{A}(\boldsymbol{r}) = \frac{\mu_0 I}{4\pi} \oint \frac{\mathbf{d}\boldsymbol{l}}{R}$$

where R is the distance from the circuit element $\mathbf{d}\boldsymbol{l}$ to P and $r \gg a$, the radius of the circle. Hence show that

$$\boldsymbol{A}(\boldsymbol{r}) = \frac{\mu_0(\boldsymbol{m} \times \boldsymbol{r})}{4\pi r^3}$$

where $\boldsymbol{m}$ is the magnetic dipole moment of the circular current I.

4 Using the definitions of $\boldsymbol{A}$ and ϕ given in (8.26) and (8.29), show that Maxwell's equations div $\boldsymbol{D} = \rho_f$ and curl $\boldsymbol{H} = \boldsymbol{j}_f + \partial \boldsymbol{D}/\partial t$ lead to the following inhomogeneous wave equations for a linear, isotropic medium of permittivity ε and permeability μ:

$$\nabla^2 \boldsymbol{A} = -\varepsilon\mu \frac{\partial^2 \boldsymbol{A}}{\partial t^2} = -\mu \boldsymbol{j}_f$$

$$\nabla^2 \phi = -\varepsilon\mu \frac{\partial^2 \phi}{\partial t^2} = -\frac{\rho_f}{\varepsilon}$$

5 Explain why the electric field due to a charge is not given by the force per unit charge, nor the gradient of its electric potential, when the charge is moving.

Chapter 9

1 Show that the displacement current density in a linearly polarized plane wave $\boldsymbol{E} = \boldsymbol{E}_0 \exp \mathrm{i}(\omega - \boldsymbol{k}\cdot\boldsymbol{r})$ is $\mathrm{i}\varepsilon_0 \omega \boldsymbol{E}$ and calculate its root mean square value when $E_0 = 5.1\ \mathrm{mV\,m^{-1}}$ and $\omega/2\pi = 1\ \mathrm{GHz}$.

2 A stationary observer is looking at a mirror that is travelling away from him at a speed $v < c$. Show that, if he shines a laser beam of frequency ν at the moving mirror, he will see a reflected frequency given by

$$\nu' = \nu\{(1 - v/c)/(1 + v/c)\}^{1/2}$$

3 Solar energy falls on the earth's surface at about $1\ \mathrm{kW\,m^{-2}}$. Estimate the r.m.s. electric and magnetic fields in the sunshine received.

4 Assuming that the earth's magnetic field is due to a dipole at its centre which produces a field of 15 mT at the North Pole, estimate the total magnetic energy in the earth's external magnetic field.

5 A radio antenna at the surface of the earth is emitting its radiation radially. If its average power is 20 kW, what is the energy flux at a domestic receiver 50 km away? What are the r.m.s. values of the $\boldsymbol{E}$ and $\boldsymbol{H}$ fields at the receiver?

Chapter 10

1 A charge q, mass m, oscillating with amplitude a at an angular frequency ω_0 radiates energy at a rate $(\mu_0 q^2 \omega_0^4 a^2)/(12\pi c)$. By solving the equation of motion of a harmonically bound charge,

subject to a damping force proportional to its speed and driven by a force $F \exp(i\omega t)$, find the mean energy of the oscillator and hence show that the damping constant which simulates the radiation is $\gamma = (2\pi\mu_0 cq^2)/(3m\lambda_0^2)$, where $\lambda_0 = 2\pi c/\omega_0$. Estimate the natural width due to this cause of a line in an atomic emission spectrum.

2 Show, from Maxwell's equations, that the wave equations for a polarizable, magnetizable dielectric of relative permittivity ε_r and relative permeability μ_r are, respectively,

$$\nabla^2 \boldsymbol{E} = \frac{1}{v^2}\frac{\partial^2 \boldsymbol{E}}{\partial t^2}, \quad \nabla^2 \boldsymbol{B} = \frac{1}{v^2}\frac{\partial^2 \boldsymbol{B}}{\partial t^2}$$

where $v = c(\mu_r \varepsilon_r)^{-1/2}$.

3 The electric vector of an electromagnetic wave in a dielectric is

$$E_x = E_0 \exp(-\beta z) \exp \mathrm{i}\omega(t - n_R z/c)$$

where β is the absorption coefficient. Show that the magnetic vector is perpendicular to E_x and find the phase difference between the vectors.

Chapter 11

1 Estimate the reflectance and transmittance at normal incidence for (a) light, and (b) radio waves, from air into water. How can you explain this microscopically?

2 Show that the reflection coefficient for radiation at normal incidence from free space on to a plane surface of material of refractive index $(n_R - in_I)$ is

$$R = \frac{(n_R - 1)^2 + n_I^2}{(n_R + 1)^2 + n_I^2}$$

For a metal at low frequencies $(\omega/2\pi)$ and conductivity γ

$$(n_R - in_I)^2 = -\mathrm{i}\gamma/(\omega\varepsilon_0)$$

Show that $R = 1 - (8\omega\varepsilon_0/\gamma)^{1/2}$ when $\gamma \gg \omega\varepsilon_0$.

3 A transparent dielectric of refractive index n has a plane boundary, forming the (y, z) plane, with free space. A linearly polarized plane wave is incident on the boundary from the medium. The magnetic field $\boldsymbol{B}$ in the incident (I), reflected (R) and transmitted (T) beams is parallel to the positive axis and has components:

$$B_{\rm I} = A \exp[\mathrm{i}\omega\{t - n(x\cos\alpha + y\sin\alpha)/c\}]$$

$$B_{\rm R} = B \exp[\mathrm{i}\omega\{t - n(-x\cos\alpha' + y\sin\alpha')/c\}]$$

$$B_{\rm T} = C \exp[\mathrm{i}\omega\{t - (x\cos\beta + y\sin\beta)/c\}]$$

Draw a diagram showing the directions of the three beams and of their electric fields, indicating the angles α, α' and β. By considering the y-dependence of $\boldsymbol{B}$ at the surface, show that $\alpha = \alpha'$ and $\sin\beta = n\sin\alpha$. Show that a solution of the equation $\sin\beta = n\sin\alpha$ can be obtained when $n\sin\alpha > 1$ by putting $\beta = \pi/2 + \mathrm{i}\delta$, where $\cosh\delta = n\sin\alpha$. If $n = 1.4$, $\alpha = 80^\circ$ and the radiation has wavelength of 400 nm in free space, find how far from the surface the magnetic field has dropped to 10% of its value at the surface. [Hint: $\sin(\gamma + \mathrm{i}\delta) = \sin\gamma\cosh\delta + \mathrm{i}\cos\gamma\sinh\delta$; $\cos(\gamma + \mathrm{i}\delta) = \cos\gamma\cosh\delta - \mathrm{i}\sin\gamma\sinh\delta$.]

4 A laser beam, having a power of 100 MW and a diameter of 1 mm, passes through a glass window of refractive index 1.59. Find the peak values of the electric and magnetic fields of the laser beam (a) in the air, and (b) in the glass.

5 Show that the wave impedance Z of a medium of permeability μ and permittivity ε is $(\mu/\varepsilon)^{1/2}$.

6 A uniform plane wave is normally incident from a medium 1 into a parallel slab of thickness l of medium 2 and emerges into medium 3 after two partial reflections. Show that there are no reflections when (a) media 1 and 3 are the same and $k_2 l = m\pi$, where m is an integer; (b) $k_2 l = \pi/2$ and $Z_2 = (Z_1 Z_3)^{1/2}$.

Chapter 12

1 Show that an electromagnetic wave with complex $\boldsymbol{E}$ and $\boldsymbol{H}$ fields given by $\boldsymbol{E} = (\boldsymbol{E}_{\rm R} + \mathrm{i}\boldsymbol{E}_{\rm I})\exp\mathrm{i}\omega t$ and $\boldsymbol{H} = (\boldsymbol{H}_{\rm R} + \mathrm{i}\boldsymbol{H}_{\rm I})\exp\mathrm{i}\omega t$, has an average Poynting vector given by $\langle\mathscr{S}\rangle = \frac{1}{2}\mathrm{Re}(\boldsymbol{E}\times\boldsymbol{H}^*)$, where $\boldsymbol{H}^*$ is the complex conjugate of $\boldsymbol{H}$.

2 A linearly polarized electromagnetic wave falls at normal incidence on a good conductor. Show from the Lorentz force $\boldsymbol{j}\times\boldsymbol{B}$ due to its magnetic vector $\boldsymbol{B}$ acting on the induced surface current $\boldsymbol{j}_{\rm f}$ in a direction normal to the surface that it produces a radiation pressure $p_{\rm r} = 2\mu_{\rm i}$, where $u_{\rm i}$ is the energy density of the incident wave.

3 Electromagnetic waves of frequency 1 MHz are incident normally on a sheet of pure copper at 0°C. (a) Calculate the depth in the copper at which the amplitude of the wave has been reduced to

half its value at the surface, if the conductivity of copper at $0°C = 6.4 \times 10^7\,S\,m^{-1}$. (b) Explain how you could calculate this depth at higher frequencies and at lower temperatures.

4 Discuss the possibility of using radio waves to communicate with a submarine submerged in seawater of conductivity $4.0\,S\,m^{-1}$.

5 A spacecraft returning to earth produces a cloud of ionized atoms of density $10^{15}\,m^{-3}$. Find (a) the plasma frequency of this cloud, (b) the cut-off wavelength for communication with the ground.

6 Show that the energy dissipated by a current I flowing in a long, straight wire of conductivity γ and radius a can be described as flowing into it radially from its surroundings. Hence show that the power dissipated per unit length $I^2/(\pi\gamma a^2)$.

Chapter 13

1 Using the Lorentz condition for the vector potential $\boldsymbol{A}$ show that the equation

$$\frac{\partial A_z}{\partial z} = \left(\frac{\mu_0 l}{4\pi}\right)\frac{\partial}{\partial z}\left\{\frac{I_0}{r}\cos\omega(t - r/c)\right\}$$

can be solved to find the scalar potential

$$\phi = \frac{l}{4\pi\varepsilon_0}\left\{\frac{\cos\theta}{r^2}q_0\sin\omega(t - r/c) + \frac{\cos\theta}{cr}I_0\cos\omega(t - c/r)\right\}$$

where l, r, z, θ are given in Fig. 13.1(b) and $I_0 = \omega q_0$.

2 For the radiation field of a Hertzian dipole the vector potential is

$$\boldsymbol{A}(\boldsymbol{r}, t) = \frac{\mu_0 l I_0}{4\pi r}\{\cos\omega(t - r/c)\}(\cos\theta\hat{\mathbf{r}} - \sin\theta\hat{\boldsymbol{\theta}})$$

and the scalar potential is given in exercise 1. Show that the electric vector is $\boldsymbol{E} = E_\theta\hat{\boldsymbol{\theta}}$, where $E_\theta = -E_0 r^{-1}\sin\omega\,(t - r/c)$ and $E_0 = (\omega l I_0 \sin\theta)/(4\pi\varepsilon_0 c^2)$.

3 The Poynting vector for the radiation field of a Hertzian dipole is

$$\mathscr{S} = \frac{E_0^2}{\mu_0 c r^2}\sin^2\omega(t - r/c)\hat{\mathbf{r}}$$

Show that the average Poynting vector over one cycle is

$$\langle\mathscr{S}\rangle = \frac{\mu_0 c l^2}{32\pi^2} I_0^2 \sin^2\theta \frac{k^2}{r^2}\hat{\mathbf{r}}$$

where I_0 is given in exercise 1 and $k = \omega/c$.

4 Show that the radiation resistance of a current-loop antenna of radius a is $20\,\pi^2(ka)^4$, where k is the wave number of the radiation. Estimate the radiation resistance for a loop with the Bohr radius a_0 emitting red light.

5 Show that the power radiated by a linear quadrupole antenna (Fig. 13.3(a)) is given by $(\mu_0\omega^6 Q_0^2)/(240\pi c^3)$ and that its radiation resistance is $4(kl)^4$. Hence show that the ratio of power radiated from a quadrupole antenna to that from a Hertzian dipole of the same length is $(kl)^2/20$.

6 The radiation field of a Hertzian dipole of moment $\boldsymbol{p}(t) = \boldsymbol{p}_0 \exp i\omega t$ has a vector potential given by

$$\boldsymbol{A}(\boldsymbol{r}, t) = \frac{\mu_0}{4\pi r}[\dot{p}](\cos\theta\hat{\mathbf{r}} - \sin\theta\hat{\boldsymbol{\theta}})$$

where $[\dot{p}]$ is the time derivative at the reduced time $(t - r/c)$ of p. Show that the radiation fields are

$$E_\theta = \frac{\sin\theta}{4\pi\varepsilon_0 c^2 r}[\ddot{p}],\ B_\psi = \frac{\mu_0 \sin\theta}{4\pi c r}[\ddot{p}]$$

and that power radiated is $P = [\ddot{p}]^2/(6\pi\varepsilon_0 c^3)$.

Chapter 14

1 For the TE_{10} mode in lossless, rectangular waveguide (Fig. 14.3(a)), obtain expressions for (a) the average electrical energy, (b) the average magnetic energy, per unit length of guide, and hence show that the total electromagnetic energy per unit length is $E_0^2\varepsilon_0 ab/4$, where E_0 is the peak amplitude of the electric vector.

2 Calculate the cut-off frequency of the following modes in a rectangular waveguide of internal dimensions 30 mm × 10 mm: $TE_{01}, TE_{10}, TM_{11}, TM_{21}$. Hence show that this waveguide will only propagate the TE_{10} mode of 10 GHz radiation.

3 A rectangular cavity has internal dimensions of 30 mm × 15 mm × 45 mm. Find the three lowest resonant modes and calculate their frequencies.

4 A microwave receiver is connected by 30 m of waveguide of internal cross-section 23 mm × 10 mm to an antenna. Find the ratio of (a) the phase velocity and (b) the signal velocity in the waveguide to that in free space, for reception at 12 GHz.

5 A rectangular cavity made from waveguide of aspect ratio 2.25:1 resonates at 8.252, 9.067 and 9.967 GHz. If these frequencies are those of adjacent TE_{10l} modes find the length of the cavity, assuming that the cut-off frequency of the TE_{10} mode of the waveguide is 6.56 GHz.

APPENDIX H

Answers to exercises

Chapter 2

1 3.10×10^{35}.
2 (a) 0 (b) $12.5\,\mathrm{V\,m^{-1}}$ (c) $25\,\mathrm{V\,m^{-1}}$ (d) $4\,\mathrm{V\,m^{-1}}$.
3 $8 \times 10^{-19}\,\mathrm{J}$.
4 $4.77 \times 10^{5}\,\mathrm{V\,m^{-1}}$.
5 $\mathrm{A^2M^{-1}L^{-3}T^4}$.
6 (a) $4.80 \times 10^{-17}\,\mathrm{J}$ (b) $1.03 \times 10^{7}\,\mathrm{m\,s^{-1}}$.
8 40 V; increased fourfold.
9 10 mm.
10 (a) Total charge of 24 nC is conserved. (b) Potentials equalized to 1.66 kV by charge flow. (c) Some energy lost to Joulean heat in wire and flash of light (spark). (d) 1.9 μJ.
11 (a) $3 \times 10^{-8}\,\mathrm{J}$ (b) $6 \times 10^{-8}\,\mathrm{J}$. In (a) $6 \times 10^{-8}\,\mathrm{J}$ work is done charging the battery, but work done on capacitor is $-3 \times 10^{-8}\,\mathrm{J}$.
12 (a) $10.8\,\mathrm{\mu C\,m^{-2}}$ (b) 600 V.
14 Force acts only where the electric field is non-uniform, at the edge of the plates over the dielectric.

Chapter 3

1 $\phi = (p\cos\theta)/(4\pi\varepsilon_0 r^2)$.
3 $(3qa^2)/(2\pi\varepsilon_0 r^4)$.
5 $E_r = E_0\cos\theta\{1 + 2(R^3/r^3)\}$; $E_\theta = -E_0\sin\theta\{1 - (R^3/r^3)\}$.
6 Lines of $\boldsymbol{D}$ are continuous and uniform within the sphere, but forced out when $\varepsilon_1 < \varepsilon_2$ and forced in when $\varepsilon_1 > \varepsilon_2$.
8 Positive at (−2, 4), (−4, −2), (2, −4), (4, 2); negative at (2, 4), (−4, 2), (−2, 4), (4, −2). No, angle must be submultiple of 2π.
10 Both at a point inverse to the centre of the dipole in the sphere.

Chapter 4

1 $\pi j_0 R^4/2$.
2 $63.7\,\Omega$.
3 8.04×10^{-3} T.
5 $8\mu_0 I/(5\sqrt{5}a)$.
6 12.8 T.
7 $1.00\,\text{N}\,\text{m}^{-1}$. Same direction, attractive; opposite direction, repulsive.
8 1.13×10^{-4} T.
9 (a) $\mu_0 Ir/(2\pi a^2)$ (b) $\mu_0 I/(2\pi r)$ (c) $\dfrac{\mu_0 I}{2\pi r}\left(\dfrac{c^2 - r^2}{c^2 - b^2}\right)$ (d) 0.
10 $9.27 \times 10^{-24}\,\text{J}\,\text{T}^{-1}$.

Chapter 5

1 (a) $B_0 S\omega \cos \omega t$ (b) $B_0 S\omega \cos 2\omega t$.
2 0.3 V.
3 37.5 MeV.
4 1.21 kA. For motion clockwise about $\boldsymbol{B}$, current flows from centre to rim.
5 1.97 kJ.
6 25.1 μH.
7 3.14 V r.m.s.
8 6.82 kV.
9 6.28 μH.

Chapter 6

1 1.96×10^{-3}.
2 1.1×10^{19} J.
3 (a) False, since they have different dimensions. (b) False; this just defines the units of $\boldsymbol{H}$. (c) False; $H = (\bar{B}_a/\mu_0) - N\bar{m}$ always has finite $\bar{m}$, although it can be very small. (d) True; $\boldsymbol{B} = \mu_0(\boldsymbol{H} + \boldsymbol{M})$. (e) False; sources of $\boldsymbol{H}$ are conduction currents and of $\boldsymbol{B}$ are conduction and magnetization currents.
4 (a) 37.6 T (b) 1.12 kA (c) 11.1 ms.
5 0.11 T.

Chapter 8

2 $E = 0.78\,\text{V}\,\text{m}^{-1}$, radial; $B = 2.6 \times 10^{-9}$ T, azimuthal.

Chapter 9

1 $200\,\mu A\,m^{-2}$.
3 $600\,V\,m^{-1}$; $2\,\mu T$.
4 $10^{19}\,J$.
5 $1.27\,\mu W\,m^{-2}$; $22\,mV\,m^{-1}$; $58\,\mu A\,m^{-1}$.

Chapter 10

1 $1.2 \times 10^{-14}\,m$.
3 $\tan\delta = n_I n_R$.

Chapter 11

1 (a) $R = 2\%$; $T = 98\%$; (b) $R = 64\%$, $T = 36\%$; interference between radiation induced and incident.
3 153 nm.
4 (a) $312\,MV\,m^{-1}$, 1.04 T; (b) $248\,MV\,m^{-1}$, 1.31 T.

Chapter 12

3 (a) 44 μm.
4 At 100 Hz, skin depth = 25 m.
5 (a) 280 MHz; (b) 1.1 m.

Chapter 13

4 $10^{-11}\,\Omega$.

Chapter 14

1 (a) $E_0^2 \varepsilon_0 ab/8$; (b) same.
2 15, 5,15.8, 18.0 GHz.
3 101, 102, 201; 6.0, 8.3, 10.5 GHz.
4 (a) 1.19, (b) 0.84.
5 119.8 mm.

Index